THE QUALITIES AND CHARACTERS OF A GOOD AND GREAT LEADER

"Character Building "

By

The Rev. Dr. François K. Akoa-Mongo

INTRODUCTION

I was born, raised and occupied some leadership positions in many Cameroonian and as well as the USA circles. At the age of 75, I have spent 50% and 50% of my life in Africa as well as in the United States. I thank my Lord for allowing me to be a part and capable to master these two surroundings. Today, because of my experiences, I am presenting this book on the kind of qualities and characters those who can save and lead Africa should have. No society or country can succeed without qualified leadership. This is what Africa in general and Cameroon in particular need right now.

According to me, this book should be ranked second after the Bible in importance as guides African leaders should hold on their two hands the Bible and this book in order to do what other developed societies have done over the years. Africans should not reinvent the wheel when the wheel was already were invented some millennia ago. I could to believe my hears when I followed a declaration of an intellectual and great Cameroonian professor who said that Africans will have to follow first all the steps in politic, economic, social and cultural development experienced by the European countries in history before they can become capable to do behave and think like Westerners. I do not believe that we Africans should seek to invent for example an electric motor when we can either buy or learn from Europeans how to make one. Should we have an African democracy, then another European democracy, for example? Democracy is democracy because of its principles and goals are the same wherever the democratic system is chosen as a system of government of the people by the people and for the people. Africa needs leaders in all her social, religious, cultural and economical organizations. Africa needs visionary, ambitious leaders, pioneers and dreamers, influential renovators, who can move wills, stimulate individual and group talents in all levels leading human beings to accomplish great projects mobilizing the forces in order to change for the best their environments. This is the only way African countries and societies could historical be numbered among the developed ones in the world.

Europeans, rather the Western world, knows the importance of the science of leadership, of management and the place that science occupies within human development; its goal is about how to bring together individual people in order to reach visions of common interests dreamed by certain or group of individuals to begin with.

On the basis of the idea and influence of one person, this one would be capable to bring others together. These projects will never be for the benefit of the dreamers who have acted as catalyzers and facilitators, but for the groups. This is what this book would like to teach and train young

Africans to lead.

Great number are projects, from African families, groups and even governments, have failed or never started just because African societies did not have good and great leaders. Since Africans have become masters of their own destiny since independence, we have gone backward almost in every area of our life compared to the colonial period. It has been that way because African leaders refuse to model ourselves after the Westerners. African leadership in general and Cameroonian in particular until today do everything according to "their own manners". A certain Pastor from the Presbyterian Church of Cameroon interrupted me, François Akoa-Mongo, in 2014 when he spoke during an annual meeting of the Municam Synod. A report was presented dealing with poor managerial skills of many agents of the EPC. This poor management concerned assets and personal. Here is what that Pastor had said to him:

"Francois, 27 years ago, you left this church and have been living in the USA; You do not know the actual realities of our society. Leave us alone to dismember our elephant according to our own manners." What that soon to retired pastor told François that day is the true definition of the philosophy of leadership in place within all circles of life in Africa. A person appointed to manage a business in Africa as well as those around him will agree that he has been given an elephant and he is going to dismember it according to his will. 99.9% of African leaders and leaders today will agree with this Pastor of the Presbyterian Church of Cameroon. If there is a book that can help, teach and train Africans in the science of leadership outside of the Bible, precisely, "

THE QUALITIES AND CHARACTERS OF A GOOD AND GREAT LEADER" should be number one. Africans, welcome this book as we are looking for the way we can succeed to develop ourselves.

Selfishness, tribalism, personal gains first, theft, paternalism, lowliness, the culture of bad attitudes, blunted consciences, embezzlement of funds and everything that makes African societies to remain primitive is caused by the spirit of bad leadership. Despite the years of African contacts with Europeans, the great number graduates from universities, titles and great executive positions, Africans will never change as long as the concept and definition of leadership in Africa will not be the same as in Europe and in other developed countries in the world.

If present and future African leaders could embody what Western experts say and teach and what I have put together in this book, if they will put into practice ideas that let the societies of the world succeed using this book, II is certain that in 25 years we could see churches, public and private institutions, business communities in all levels and sectors, agencies, industrial groups, and everything about employment in Africa will become a successful story.

My prayer is that the Lord use, besides the 40 more books I have written and published and this book on leadership for the recovery of our African societies that is becoming new Haiti instead of modeling themselves after the West.

To God be the glory.

His slave, François K. Akoa-Mongo

Machiasport, on November 30, 2016

DEDICATION

Thinking about to whom should I dedicate this book
took some moment, and I became convinced that
my nephew, Francis Akoa should be the one.
Francis is one of the few Cameroonians who have
demonstrated great leadership in management
skills. I am convinced that he understands more than
any other the need of such a book in leadership in
our African societies, which is bringing many of us to
the point of losing any hope of a better future for
next generations. Only leaders who follow laws,
rules, attitudes, and who incarnate and put into
practice leadership qualities and characters that will
save Africa and make our continent the cradle of
new hope in the world.

CONCERNING THE AUTHTER

The Rev. Dr. François K. Akoa-Mongo has been a Pastor within the Holy Ministry of Jesus Christ since 1967. He is a man who lives and knows African and American cultures. Born in Cameroon, he pursued his education in both countries. He also worked as a teacher and Pastor in both societies. He was ordained within the Presbyterian Church of Cameroon.

He has a Master Degree in Theology from Bangor Theological Seminary, USA, Maine, a Master's in Teaching Foreign Languages, and a Ph.D. in Education from the University of Maine in Orono.

He is the author of over 42

Christian books that are in English and French. They can be bought in "Amazon.com"

He resides with his wife Kathy in Machiasport, where he has

been a Senior pastor for the last 25 years serving
the congregational parish of Machiasport. All his 9
children live
in the USA.

TABLE OF CONTENT

CHAPTER I

LEADERSHIP DEFINITION:

To influence and bring people together.

Leadership is the process by which a person influences a group of people to achieve a common goal (Peter NORTHOUSE, Leadership - Theory and Practice, Sage). Being a leader is a recognition, not a status.

There are many definitions of leadership, some very broad, others narrower. I propose a simple definition of leadership, which nevertheless covers the essential part of this notion.

Leadership is: 1. A personal ability to influence and bring a group together;

a / - In order to achieve a common goal,

b / - within a mutual trusting relationship,

c / - and for a limited time.

THE 4 CHARACTERISTICS OF THE LEADERSHIP:

1. A leader is someone who influence individuals and brings the group together. Leadership is an influential authority, based on the relationships with the members of a group. To achieve this goal, this involves the following actions

A / - He must effectively communicate with group members;

B / - He must make the team adhere to a common goal;

C / - He must motivate team members to achieve

common goals. Right from the start, one is identified as a leader because he has an objective that is for the common good; It is that objective one can't accomplish alone, but influences others who now adhere to his ideas.

2. In order to achieve a common goal, the group is considered as a group because his members put together their efforts and talents to achieve a common goal. Any influence from a leader which is not conform to this definition is not what we believe to constitute a good and true leadership according to this book. Concerning leadership: Ours is manifested in three levels: a - there must have a vision, which inspires the members of

the team. This vision is what give meaning to the action; b /

- this vision has one or more goals; these goal or goals

become aim of all actions of the group. These goal or goals

constitute the **"putting into words"** of a vision; c / - This goal or goals should have several objectives, which are strategic and operational. All objectives must be translated into measurable and organized according to some indicators over time period. A leader derives his authority from the members of the group; in order terms, the members of the group delegate their authority to the leader in order to guide them toward the fulfillment of the vision that is now common to the leader and the group.

3. Establishing relationships of mutual trust:
The group recognizes the leader because of his positive influence toward goals sets. This implies the following two points: group confidence towards the leader and the leader confidence towards the group;

a / - In this way, there is mutual respect and reciprocal understanding between leadership and group members; b / -
In action, the leadership is exemplary; the rest of the group
follows the leader.

4. Leadership is temporary: One cannot be a leader at
any time and for a long period of time, because the environment plays a preponderant role in leadership (a crisis situation for example can radically upset the leadership within a group). It is up to the leader to be able - when the situation demands it - **to "let go"** of his leadership peacefully; which implies :
a / - Leave the levers of participatory leadership (solicit group members and share responsibility for decision-making) and delegated leadership (transfer responsibility for decision-making);
b / - Be able to leave room for others when

necessary (especially when a person is more competent on a given subject);

c/ - Be humble, because a leader exists only through his team.

So if the group trusts someone else who has the visions that promises better the common goal(s).

CHAPTER 2

COMPREHENSIVE STUDY OF SOME QUOTES ON LEADERSHIP.

Before we can immerse ourselves in the study of Leader and Leadership, I chose some 19 quotes from others about their understanding of what is leadership. After each quotation, I have prepared some questions to stimulate good assimilation of each one. So, let's spend time, whether alone or working in groups, to first master all phases of the questionnaire.

Leadership is a kind of power to empower others to become involved in a collective project or to achieve a common goal together. This power cannot be decreed, it is conferred on the leader by his natural authority, his confidence in others, his experience, and so on. It is because others consider him as a leader that has the power to make them act. So if you also want to become a true leader, first learn first these lessons learned and inspired from quotes of eminent leaders.

1. Jean-François Rial : "LEADERSHIP IS THE PRODUCT OF AN ATTACHING PERSONALITY AND ASSOCIATED WITH A FORCE OF CONVICTION THAT ALLOW OTHERS TO BE A PART ON AMBITIOUS PROJECTS. "

A-What is the relationship between leadership, its belief and the current project?
B- Why is the idea of training others important in leadership?

2. Dwight Eisenhower: "LEADERSHIP: THIS IS THE ART OF SHOWING SOMEONE THINGS TO BE

DONE, BECAUSE HE WANTS TO DO THEM. "

A- What does Eisenhower mean by "leadership is an art?" Can everyone be a good and great leader if not an art? B- Why does someone else have to do what you want to see done? And why is it necessary that he should want to do it?

3. Nicolas De Tavernost: LEADERSHIP IS TRUSTING AND HAVING CONFIDENCE AROUND ONESELF.

A- What is the importance of the leader trusting himself?

B- Find other synonyms of trusting.

C- Why a leader should have confidence on those around him?

4. Françoise Gri : "LEADERSHIP TRANSLATES THE CAPACITY OF A LEADER TO GAIN FROM TEAMS A STRONG AND SUSTAINABLE ADHESION

IN ORDER TO REALIZE AN AMBITIOUS PROJECT."

A- Why does the leader need teams with a strong and sustainable adhesive ideas ?

B- Are these strong adhesive ideas based on his person or gift of leadership ?

C- Why such adhesive ideas dangerous if based on personality ?

D- Find some examples of these two kinds of strong attachments.

5. Pierre Vareille : "LEADERSHIP AND MANAGEMENT CONTRIBUTE TOGETHER TO DEVELOP AND EXECUTE A CERTAIN STRATEGY. BUT ONLY THE LEADERSHIP DEFINES VISIONS AND FRAMEWORK OF THE CORPORATE STRATEGIC ORIENTATIONS. "

A- Is it really possible to separate leadership and management in implementing a certain strategy? Why and why not?

B-Why is management should not concerned with the vision of the company?

C-What is the meaning of the word "strategies" ? Give some examples.

D-Can we talk about a true business without strategies?

6. Guillaume Poitrinal : "WHEN THE LEADERSHIP IS PLAYED ON THE DOUBLE PARTITION OF REASON AND EMOTION, IT IS A POWERFUL MOTOR THAT GIVES MEN AND WOMEN THE ENERGY NEEDED TO EXCEED THE LIMITS OF THE POSSIBLE. "

A- Why does a good leader have to play both reason and emotions to become a powerful engine in the business venture?

B- What is the relationship between reason, the emotions, and energy of the group necessary to go beyond the limits of the possible? (Motivation!) C- Define the word "motivation"; give some examples through which people will be motivated to go beyond the limits of what is possible.

7. Georges Pauget : "LEADERSHIP IS, AT THE TIME, TO BE BEFORE AND GOING FORWARD. "

A - Explain this definition of leadership," being \ ahead and moving forward ".

B - Can a leader be ahead without going forward? Why and why not?

8. Napoleon Bonaparte : A leader is a merchant of hope.

A-What is the place of hope in relation to success in business?

B-What are the similarities between a merchant and a leader?

9. Peter F. Drucker : Management is doing good things, leadership is doing the right things.

A- Explain in other words what Peter Drucker states in these terms.

B-Is it possible that managers take the place of the leader and vis-versa?
Why and why not?

10. George S. Patton: Do not tell people how to do things, tell them what to do and let them surprise you with the results.

A- What is the danger that the leader tells people how

B- to do things ?Top of Form9. Peter F. Drucker Management is doing well things, leadership is doing the right things.A- Explain in other words what Peter Drucker states in these terms.

C-Is it possible that managers take the place of the leader and vis-versa? Why and why not?

10. George S. Patton : Do not tell people how to do things, tell them what to do and let them surprise you with their results. A- What is the danger of telling people how to do things as a leader?

11. Peter F. Drucker : Management is doing well things, leadership is doing the right things.
A- Explain in other words what Peter Drucker states in these terms. B-Is it possible that the manager takes the place of the leader and vis-versa? Why and why not?

12. George S. Patton : Don't tell people how to do things, tell them what to do and let them surprise you with their results.
A- What is the danger of telling people how to do things as a leader?

11. Jesus Christ : If a blind man leads a blind man, they will both fall into the ditch.

A-In what way can a leader be considered blind ? Give some examples of a blind leadership.

13. Sam Rayburn : You can't be a leader, and ask others to follow you unless you know how to follow others.

A- In what way a good leader does exactly what Sam Rayburn says?

B- If he doesn't , what would be the consequences of a such leadership?

14. Someone stood at Andrew Carnegie's tomb and said, Here lies a man who knew how to appeal to the service of better men than himself.

A- According to this statement, how important is the choice of members of the team by a collaboration's leader ?

B- What is the importance of the phrase "knowing

how to appeal"?

15. Henry Gilmer : Look over your shoulder from time to time to make sure someone follows you.
A- Name some possible consequences to the project when a leader does not follow this advice.
B-Name some reasons a leader should continually apply this advice.

16. Manfred Kets de Vries : The strength to articulate a harmonious and pragmatic vision is closely related to the arts of communication, persuasion and the team coalition under the leader.
A- Name the four necessary elements that define good leadership here. Explain.
B-What is the relationship between vision and change for the better?

17. Kouzes and Posner : Leadership is the art of mobilizing others to want to fight for common aspirations.

A-Name at least three elements required in this sentence, which are lacking in the majority of African leadership. Demonstrate the reality of this conclusion.

18. Rost : Leadership is a relationship of influence between leaders and followers who wish real changes that reflect their mutual needs.
A-Can we be qualified as a good and great leader if a certain percentage of egoism, tribalism, sexism and bias is in us ? Why and why not?

19. Kotter : Leadership is a process of producing changes, setting direction, aligning and motivating people's efforts and talents for given goals.
A-What is the importance of monitoring the process, determining the direction and motivating the teams?

20. R. J. House : Leadership is the ability of an individual to influence, motivate and enable others to contribute to the effectiveness and success of an organization of which they are members.

A- Why is the contribution of others necessary for the effectiveness and success of an organization of which you are members?

B-Find the equivalents of "influencing" in this definition.

II - DUTIES:

A- After covering this first part of these quotes and dealing with the questionnaire gaining good understanding of each one quote, compose your personal two or three quotes describing your own understanding of what is a good leadership.

B- To what extent these quotes have opened your eyes to better understand the high expectations any society, or organizations should place on those in charge of the leadership today?

C- Based on this chapter, talk about the flaws and damages of poor leadership.

CHAPTER 3

HOW TO RECOGNIZE A LEADER

Whenever we meet with others, it is easy to recognize the one who embodies the gift of leadership. The following qualities will be discovered in him or her. We will enunciate aspects, values, attitudes, characters, and ways of life describing that person.

1-One who is destined to become a good leader has the following gifts: to persuade, convince, influence, inspire, guide and lead others towards a common goal. It is not only a matter of persuading others, but of making them partners for a common cause. There are people who know how to persuade and convince others for personal, selfish reasons and most often for non-progressive causes. The type and kind of leader we

have in mind build societies, brings together others of goodwill to mark history. This kind does not think of self, but works for the common cause, for what will make a difference in the lives of many.

2- Whoever may become a good leader uses the following arts : to mobilize and obtain the voluntary adhesion and determination of the other members of the group. He controls some specific strengths, traits and qualities that are innate and personal. One can go to school, read the books and associate himself with those who have these strengths, traits and qualifications; but it is difficult to acquire perfectly these arts. Leadership is innate; it can only be developed and not acquired. According to the book written by W. Bennis and published in 1991, titled "Profession leader", these specific strengths, traits and qualifications are personally dominant in it.

3. He who has the gift to become a leader has and lives having a certain passion. For those who know him and have lived around him for a long period of
time, sometimes from childhood, they will tell you about this specific passion in a certain perspective of life. One may even think of that passion as his calling, his vocation to become a professional. When he starts to share, explain his passion and motivation to others, he incarnates the ability to transmit others his inspirations, views and hopes. He makes others feel what he feels, to see the destiny of his imagination. He knows how to communicate the enthusiasm that lies within him by choosing convincing words and expressions.
Whoever is born to become a leader has the gift of selling his ideas and to receive the support of all

who listen to him.

4. Whoever is destined to become a leader knows the value of integrity, because it is what he practices every day. If you are looking for someone who is integer wherever you life today, the name of that person will come out in your mind. You see, one of the most valuable and necessary qualities any leader should have, not something he learned but which is inborn, that has built his conscience is integrity, followed by sincerity, truth and the love of social justice. You can't have integrity without the other named qualities.

5. The one who is destined to become a leader already knows the value of perseverance, paying attention to others, and able to work with others [...] by telling the truth.

6. Whoever becomes a leader has accumulated the experience that helps him to make good decisions and the ability to quickly identify the issues.

7. The one who has the tendency to become a leader knows how to win the trust of others. The confidence is not given but acquired, gain more and more. This is
what a good leader does.

8. Before becoming a leader, one must be curious, audacious and courageous. A leader is someone who knows how to ask right questions and seek understanding in life. He is a monster of creativity, vision and initiative. He is not afraid to think outside the box, to find new and innovative solutions. Through this trait of character, he become the envy and admiration of the others who seek to follow him in his success.

9. He who becomes a leader already has that status in him, in his nature and daily life, he acts with that charisma in many contexts. This quality alone is not always sufficient. The complexity of human relations requires that the leader be all the more integral and visionary and less charismatic.

10. Here is a very important point about leadership: Leadership is not a power acquired through hierarchical structure within an organization; It is a catalytic position between different elements as actors in a social organization. The manager is not necessarily a leader; the latter can lead his team towards fixed objectives. Only the leader is needed to inspire teams to ensure that goals of the organization are voluntarily exceeded. Moreover, in the near future, managers will be led more to be true leaders if they want to reach the

high level of growth for their organization or the progress of their career.

We have just covered the ten qualities or unborn traits that are most often found in people who may become leaders: a passion for something, integrity in their daily lives, a persistent in their visions, someone who pays attention to others, an experienced man empowered to make good decisions, capable of winning the trust of others and a charming man in many ways. The following qualities are not the least: one can not become a leader and not being curious, audacious and courageous. These qualities remain unborn, unlearned in all good and great leaders in all the societies of the world.

THE THREE MOST IMPORTANT QUALITIES OF A PERFORMING LEADER: Here are the most important three qualities of a successful leader:

(1) He helps people to take responsibility in achieving a common goal in a group.

(2) To be a leader doesn't mean to be a boss, a parent, a coach, a politician, or someone in any social position.   The true role of a leader is to guide others to assume their roles.

Finally, (3) an effective leader knows how to get people to excel in order to get maximum results under a minimum of supervision.

From these three qualities concerning a leader, we can draw the following conclusions:

A. The nature and place of the leader in a given body is that of a facilitator. He is a sub-actor, the lubricant oil in the machine and the mind in the body.

B. Titles such as patron, boss and others used in hierarchies do not make sense when speaking about a good leader. His place within the organization is recognized in the motivation and the performance of the group beyond expectations.

C. Responsibility transforms the leader to be a good communicator.

CHAPTER 4

A LEADER IS A PROMOTOR OF MANAGERIAL ACTIVITIES

The leader leads an employee to become responsible by taking all the necessary time explaining in depth and in several ways reasons for his expectations and subsequent actions, of possible decisions which will be made by the managerial team and the basis or cause of which such pronouncements, must be communicated in a timely manner. Such a training gives tools and appropriate knowledge helping an employee to act without constant recourse to the boss when an unusual situation occurs. Activity-based management eliminates subordination, develops responsible personalities, and promises personal and collective performance.

A successful leader is the one that helps build trust within the team that becomes more and more autonomous and able to choose the best options according to desired collective results.

Unfortunately, there are still many business leaders who indirectly ask employees to leave their brain at the door before entering the work place and take it back on their way out. This latter approach is the "results management" approach , because the employee becomes a robot and not a responsible person.

Here are in more detail, the three most important qualities which are required of a such leader to achieve this idea of accountability:

1. **He must know how to clearly establish and communicate the objectives and the reasons of assignments:** It is its job to clearly communicate the objectives and reasons of the company to the employees who should reach them

as results. If you want people to follow you, you need to know where you want to go. Goals set should be easy to understand and measurable so that people under your responsibility can assess them and know where they

are at any point in the process. After communicating your goals, you must identify and communicate the reasons why you want to achieve these goals. One way to do this is to make a list of the five most important reasons why you and your team must reach these goals. The clearer your explain, the stronger your reasons, the greater your chances of reaching your target.

2. A good leader uses retroactive communication: It is not what you say which is important, it is what people understand and retain in your message, which is very important. Many leaders take for granted the means they communicate their ideas and very often only once, or twice. This should not automatically implies that the team has understood the message and that all members will act according to what you have communicated to them . It is your duty to check the level of understanding in the people under your direction and to get there you expect. Well, good leaders question a lot and reiterate their message by making sure that each time it is presented in a different way, until the understanding of the interlocutor is well understood and understood at 100 %.

3. He is a model by his actions: The performing leader
is an example not only by what he says, but especially what and the way he puts it in actions, because his actions produce a much stronger effect than what he says. When words and actions of the leader are not in harmony, greater will be difficulties he will encounter. People are reluctant to follow a leader who asks others to do things that he can't do

himself. The values you believe in and which you have established must first be lived by you, otherwise you will be considered an impostor. If your actions do not position as examples to follow. Know that you engage in an endless battle with your subordinates. Your role as a leader remains the cornerstone of success in your company, in your family or team. In fact, your ability to make people work in a way that works well must bring group synergy towards a common goal.

Becoming an excellent leader is a daily and ongoing learning. Even if some people have better predispositions to become a good leader, you are not born a leader, you become one. So apply each day to improve the three qualities mentioned above and, surely, sooner or later, you will get exceptional results from your team.

CHAPTER 5

TEN USEFUL QUALITIES NEEDED TO BECOME A   LEADER

In recent years, one of the most notable investigations in the field of modern psychology has been to study the characteristics of leaders. The least to be said is that today, more than ever, being a good leader implies intrinsic qualities without which it would be almost impossible to run a business, an institution or an organization.

HERE ARE SOME CHARACTERISTICS AN IDEAL LEADER MUST HAVE. These qualities do not necessarily describe leaders, although most of them are unavoidable for those who run for the first time a modern globalized market.

1. Organizations, especially international organizations, **require leaders who can effectively adjust their leadership to work in different environments.** Most of the headquarters found in the United States are cross-cultural because of the different cultures that live and work there. One can facilitate the position of leadership in depending on measures and outcomes - not even on skill, for it takes many skills to master. The effectiveness of a group is directly linked to the effectiveness of process. If the group is in good working order, the facilitating leader will have a light hand on the process. On the other hand, if the group is not very functional, the facilitator leader will have more guidance to help the group carry out its process. **Effective facilitating leadership involves monitoring and dynamiting the group, offering process suggestions and intervening to help the group stay on track.**

2. Leadership laissez-faire gives authority to employees. According to this kind of leadership, subordinates are allowed to work as they wish with minimal interference or none from the leader. This kind of leadership has been consistently found to be the least satisfying and less effective management style.

3. Transactional Leadership. It is a leadership that keeps or continues the status quo. It is also the kind of leadership that involves a process of exchange, through which the followers obtain immediate and tangible rewards for the realization of the leader's orders. Transactional leadership may seem rather basic, with its emphasis on exchange. Be clear, focus on expectations, giving feedback are all important leadership skills. Transactional leadership behaviors may include: (a) *clarifying what is expected* of the performance of the followers; (B) *Explaining how to meet these expectations;*

(C) *Awarding rewards which are conditional* upon the attainment of the objectives.

4. Coaching Leadership: Coaching leadership is a teacher and supervisor follower style. A coaching manager is very operational in the context where results / performances require improvement. Basically, in this kind of leadership, followers are helped to improve their skills. Coaching leadership: motivates inspires, and encourages followers.

5. Charismatic Leadership. The charismatic leader manifests his revolutionary power. The charisma does not mean a change of behavior; it really involves a transformation of the values and beliefs of the followers.
Thus, we can distinguish a charismatic leader from a merely populist leader who can affect attitudes toward specific objects but who is not prepared as the charismatic leader to transform the underlying

normative orientation that structures specific attitudes.

6. Visionary Leadership. This form of leadership involves leaders who recognize that leadership methods, steps and processes are all gained from and through people. Most great and successful leaders have the aspects of vision in them. However, those who are very visionary are those who are seen as exhibiting visionary leadership. *Exceptional leaders will always transform their visions into realities.*

7. Task-based Leaders. Task-oriented leaders focus only on the work accomplished and can be autocratic. They actively define the work and roles required, set up structures and plan, organize and monitor work. These executives also perform other key tasks, such as creating and maintaining performance standards. However, *because task-*

oriented leaders do not tend to think much about the well-being of their team, this approach can suffer from many gaps in autocratic leadership, including motivation and retention problems.

8. Bureaucratic leaders: Bureaucratic leaders work "by the book". They strictly follow the rules and ensure that their people follow the procedures accurately. *This style of leadership follows a narrow set of standards.* Everything is done in an exact and specific way to ensure safety and / or accuracy.

10 CHARACTERISTICS OF CHARISTMATIC LEADERS

1. It has the vision:

A good leader is above all a good visionary. He digs underneath appearances and finds new and innovative ideas. He has the ability to see the world of the future. This quality implies good intuition and a determination to do things differently.

2. Vision on mission

Being a good visionary is one thing, but also one should *be able to translate his visions into missions.* The leader knows exactly what his mission is. He knows why his company, his group or his institution exists, and he sets clear and precise objectives to achieve.

Objectives so clear that he has no difficulty in transmitting them easily to his team in order to federate around the mission he has set himself.

3. He has passion / motivation:

A good leader is bound to be an enthusiast. He loves what he does and knows that is *"his raison d'être"*. He works with enthusiasm and is a motivated man. Almost all the leaders studied by psychologists have this point in common: they are men *"obsessed"* and extremely focused on their mission.

4. It has the ability to make good decisions: Making decisions is the most common thing in the leader's daily life. Whether it is a decision of great or small size, the real leader knows how to show a great sense of responsibility and knows how to assume his choices. He does not allow doubt to weigh on him and make him deviate from his convictions. *"I DO NOT REMEMBER THE FACT THAT MY DECISIONS ARE GOOD OR BAD. WHEN I TAKE A DECISION AND AM CONVINCED, I ARRANGE FOR IT TO BE THE GOOD '* Mohamed Ali.

5. He has a perseverance: This is the first characteristic of determined men. It is the ability to get to the end of his ideas and to bring them to fruition. The perfect example to illustrate the persistence is an anecdote relative to Henry Ford, the creator of the FORD brand, which is so far one of the most important in the world of the automobile. In the early 1930s, Henry Ford had the idea of producing a V-8 engine (8 cylinders in one block) designed to be the most economical on the market . This implied very unlikely conditions, but he immediately asked his engineers to conceive it. The latter were categorical and unanimous: "This engine is impossible to conceive." Ford replied, "Do it anyway." After several months of reflection, the engineers were still unable to do so. They leave again to see Ford for an umpteenth time of the impossibility of conceiving the engine. "Continue," he said. "I want it, and I'll have it." They resumed their work, and a few weeks later, as if by magic, they discovered the secret of building the most economical V8 engine. This is how the famous Ford V-8 engine was born, through the fruit of the perseverance of Henry Ford.

6. He has self-confidence: No one can arouse the confidence of others unless he has it in himself.

Self-confidence is an important quality that any good charismatic leader must possess. It is about believing in one's potential and ability to achieve one's goals, whatever the obstacles that stand in the way. To have confidence in oneself is to be a convinced man, to have a strong vision and the will to see it fulfilled; Is to say "it is possible" when everyone claims the opposite.

7. He has a great presence of mind: It is an indispensable capability for the good leader. He allows it to anticipate the most critical situations. He involves great attention and an alert state of mind. The leader sees the danger from a distance and prepares for it in time.

8. He has empathy: It means knowing how to put oneself in the place of others and be receptive to their impressions. *When those who follow you know that you are understanding and open, they will be* *more motivated and more willing to collaborate with you and share your vision.*

9. He has great emotional intelligence: It is the faculty to control one's emotions, to grasp the tension and the pressure of everyday life. The charismatic leader has the wonderful ability to keep his composure and lucidity, whatever the situation in which he finds himself. To be emotionally intelligent is also to know how to transmit this state of mind to its citizens.

10. He is modest: Leaders; the real ones ; Those who know how to create enthusiasm in others are people of great modesty. Whatever their degree of height on the social scale, they know how to be accessible, and their relations with others are most natural. There is no such thing as forcing respect, while maintaining a good atmosphere of collaboration.

CHAPTER 6

THE FIFTEEN (15) DIFFERENT LEADERSHIP STYLES

Leadership style is the way a person uses the power to direct other people. There are variety of leadership styles based on the number of followers. *Most appropriate leadership style depends on the function of the leader, the followers and the situation.*

Some executives can't work comfortably with a high degree of employee involvement or subjects in decision making. Some employers do not have the capacity or the desire to take responsibility. In addition, the specific situation helps to determine the most effective style of interactions. Sometimes leaders need to deal with problems that require immediate solutions without consulting topics.

We will cover 12 different types of ways people tend to lead organizations or other people.

Not all these styles would be judged suitable for all kinds of situations, you can study them and see which one fits right to your business or situation.

1. Autocratic Leadership:

The autocratic leadership style is centered on the boss. In this leadership, the leader holds all authority and responsibility. In this leadership, leaders make their own decisions without consulting their subordinates. They make decisions, communicate them to subordinates and expect rapid implementation. The autocratic work environment normally has little or no flexibility. In this kind of leadership, guidelines, procedures and policies are all natural additions to an autocratic leader. *Statistically, there are very few situations that can really support autocratic leadership.* Here are some of the leaders who support this kind of leadership: Albert J Dunlap (Sunbeam Corporation) and Donald Trump

(Trump Organization) among others.

2. Democratic Leadership: In this leadership style, subordinates are involved in decision-making. Unlike autocracy, this direction is centered on the contributions of subordinates. *The democratic leader assumes the ultimate responsibility, but he is known to delegate authority to other people, who determine the work plans.*

The most unique feature of this leadership is *that communication is active up and down.* As far as statistics are concerned, democratic leadership is one of the most privileged leaders, and it implies the following: *fairness, competence, creativity, courage, intelligence and honesty.*

3. Strategic leadership style: Strategic leadership is one that involves a leader who is essentially *the leader of an organization.* The strategic leader is not limited to those who are at the top of the organization. He caters to a wider audience at all levels who want to create a high performance life, team or organization.
The strategic leader bridges the gap between the need for a new possibility and the need for practicality by providing a prescriptive set of habits. Effective strategic leadership delivers assets based on what an organization naturally expects from its leadership in times of change. 55% of this leadership normally involves strategic thinking.

4. Transformational Leadership: Unlike other styles of leadership, transformational leadership is all about initiating change in organizations, groups, oneself and others. Transformational leaders motivate others to do more than originally planned and often even more than they thought possible. They set higher expectations and generally achieve higher performance.

Statistically, transformational leadership tends to have more committed and satisfied followers. This is mainly due to the fact that transformational leaders empower the followers.

5. Team Leadership: Team Leadership involves creating a living image of his future, where he stands and what he will represent. Vision inspires and provides a strong sense of purpose and direction.

Team leadership is about working with the hearts and minds of all those involved. It also recognizes that teamwork may not always involve trust in cooperative relationships. The most difficult aspect of this leadership is whether or not it will succeed. According to Harvard Business Review, team leadership may fail due to poor leadership qualities.

6. Intercultural Leadership : This form of leadership normally exists where there are different cultures in society. This leadership has also developed as a way to recognize the first riders working on the contemporary globalized market. Organizations, especially international organizations, require leaders who can effectively adjust their leadership to work in different environments. Most of the directions observed in the United States are cross-cultural because of the different cultures that live there and work there.

7. Leadership Facilitation : Leadership facilitation is too dependent on measures and outcomes - not a skill, even if it takes a lot of skills to master. The effectiveness of a group is directly related to the effectiveness of its process. If the group is in good working order, the facilitation leader uses a light hand on the process.

On the other hand, if the group is not very functional, the facilitator leader will have more guidance to help the group carry out its process. Effective facilitation leadership involves monitoring group dynamics, offering process suggestions and interventions to help the group stay on track.

8. Leadership Leave : Leadership laissez-faire gives authority to employees. According to AZ central, departments or subordinates are allowed to work as they wish with minimal interference or none. According to research, this kind of leadership has been consistently found to be the least satisfactory and less effective management style.

9. Transactional Leadership : It is a leadership that maintains or continues the status quo. It is also leadership that implies a process of exchange, through which the followers obtain rewards Immediate and tangible for the realization of the orders of the leader. Transactional leadership may seemBasic, with its emphasis on exchange.

Be clear, focus on expectations, give feedback are all leadership skills important. According to Boundless.com, transactional leadership behaviors may include: (a) clarifying what is expected of the performance of the followers; (B) Explain how to meet these expectations; (C) To award rewards which are conditional upon the attainment of the objectives.

10. Coaching Leadership : Coaching leadership

is to teach and supervise followers. A coaching manager is very Operational in the context where results / performance require improvement. Basically, in this kind of leadership, followers are helped to improve their skills. Coaching leadership does the following:

Motivates followers, inspires adepts and encourages followers.

11. Charismatic Leadership : In this direction, the charismatic leader manifests his revolutionary power. The charisma does not mean a change of behavior. It really

Transformation of the values and beliefs of the followers.

Thus, it distinguishes a charismatic leader from a leader Simply populist, which can affect attitudes towardSpecific objects, but which is not prepared as the Charismatic leader is to transform the underlying normative orientation that structures specific attitudes.

12. Visionary Leadership : This form of leadership involves leaders who recognize that the methods, steps and leadership processes are all obtained from and through people. Most great and successful leaders have the aspects of vision in them.

However, those who are very visionary are those who are seen as exhibiting visionary leadership. Exceptional leaders will always transform their visions into realities.

13. Task Leaders : Task-oriented leaders focus only on the work accomplished and can be autocratic. They actively define the work and roles required,

Structures, and plan, organize and monitor the job. These executives also perform other key tasks, such as creating and maintaining performance standards.

However, because task-oriented leaders do not tend to think much about the well-being of their team, this approach can suffer from many gaps in autocratic leadership, including motivation and retention problems.

14. Bureaucratic Leadership : Bureaucratic leaders work "by the book". They strictly follow the rules and ensure that their people follow the procedures accurately. This style of Leadership follows a narrow set of standards. Everything is done in an exact and specific way to ensure safety and / or accuracy. You will often find this leadership role in a situation where the working environment is in favor, and sets of specific procedures are necessary to ensure safety. A natural bureaucratic leader will tend to create detailed instructions for the other members of a group.

The disadvantage of this leadership style is that it is ineffective in teams and organizations that rely on On flexibility, creativity or innovation.

15. Servant leaders : Servant leaders often lead by example. They have a high integrity and lead with generosity. In many ways, servant leadership is a form of democratic leadership because the whole team tends to be involved in decision-making. However, servant leaders often "drive from behind", preferring to stay away from the limelight and letting their team accept recognition for their hard work.
However, other people believe that in situations of competitive leadership, people who exercise servant leadership may end up left by leaders using other leadership styles. This leadership style also takes time to apply correctly: it is ill-suited in situations where you have to make quick decisions or meet tight deadlines.

CHAPTER 7

THE 6 STYLES OF LEADERSHIP

Leadership is a combination of personal qualities and interpersonal skills. Daniel Goleman highlights 6 styles of leadership in the Harvard Business Review: the executive, the leader, the visionary leader, the collaborative leader, the participative leader and the leader they call "coach".

1. Direct Leadership (Coercive) : As Blanchard points out in his theory of situational management, directive leadership is the most authoritarian. It is less relational-based than the other 5 types of leadership.
It imposes as a slogan "Do what I tell you to do" so it commands directives. This style of management has the advantage of developing rapid and concrete

advances. But if it is not used wisely, it can generate passive resistance among its employees and prove to be counterproductive. Often set up in the event of a major crisis, directive leadership makes it possible to take a tight turn in a particular situation, with some people. The manager needs to be the only decision maker.

2. Leadership as a Leader (Pacesetting) : No less demanding that the leader, the leader expects his collaborators excellence. It is he who, with a suspicion of authority, sets the example to follow, he gives the rhythm. For his team, he wants to achieve the highest level of performance.
He is like his directive cousin, more focused on objectives rather than on an overall vision. Being less attentive about the Human, this type of management sometimes has negative repercussions because it removes those who do not follow the leader. If the entire team manages to follow the

pace imposed by the leader, the objectives are effectively achieved. To have a maximum chance of being followed, the leader must take his modeling role seriously. By showing the example it stimulates the motivation of its employees.

3. Visionary Leadership (Authoritative):

It is a charismatic leader who federates his teams around an ideology, a vision. It gives the why, the sense of this vision but leaves it to its managers to take care of how to carry out the realization.

His asset? Communication. He knows how to talk to his teams, how to reboot their energy. It is nevertheless necessary to be careful not to be too evasive in his way of passing the vision. It must be directly applicable! As a true speaker, the visionary leader brings meaning to changes, he knows how to federate around his objectives.

4. Collaborative leadership (Affiliative): It

essentially seeks cohesion, expertise and harmony. It leaves plenty to exchange to understand the needs of its teams, identify problems quickly. This vision reinforces the motivation and confidence of employees in the short term (even in the medium term).

The collaborative leader quickly encounters boundaries with high-performing collaborators who expect a more demanding model like the Lead Partner. The collaborative leader only values the group and does not allow individuals to flourish individually.

5. Participatory Leadership (Democratic) : He advocates collective intelligence in the first place. He lets all the members of his team participate, take initiatives. For the participatory leader dialogue is the key to the good idea. In this climate, everyone feels heard and above all listened to. This style enhances creativity through brainstorming. It is efficient on a rather long term.

In the quest for change, this leadership style can be effective. It makes it possible to obtain democratically the commitment of its collaborators.

6. The leader "coach" (Coaching): It aims at the autonomy of each member of its team and helps them to develop skills. He believes in the potential of each of his employees. He invites his team members to be visionary to themselves.

It is an ambivalent management, it is about giving the keys to success while leaving autonomy. There are several difficulties for the coach. It must cultivate collective intelligence while being careful not to neglect its own objectives. It must set fair, defined and effective limits. It needs to start constructive feedback. This type of leadership applies to a long-term project, it is not efficient when it comes to reporting results quickly.

CHAPTER 8

WHO ARE THOSE WHO MARK THE STORY ?

The Chemistry of Business Through the Ages:

There are many ways to leave your mark in the world, no matter your type of business chemistry. The four following historical figures have left their mark in the history of the world because they knew how the chemistry of business they incarnated.

1. **Queen Victoria:** She embodied the type of Guardian leader. Queen Victoria was the British monarch who had the longest reign in history. She was the advocate of all that proved and that was true. It is because of this that an entire era was

named after her! With its style based on the meticulous control principle, Victoria is a great example of the guardian type.

2. "Great events leave me alone and calm; These are mere trifles which irritate my nerves. - Queen Victoria

3. President Theodore Roosevelt: He embodied the type of leader PilotTheodore Roosevelt was the youngest American president ever sworn in at the age of 42 and the Win the Nobel Peace Prize. His care, experimental and never attempted, was of the type of a pilot: If he met, whether it was an asthmatic problem of his own child, a political feat or a question of genius.
For the construction of the Panama Canal, he set to work to finally overcome this obstacle. As President, he was shot during a campaign and even continued

to deliver a 90-minute speech. For him, after all, this
Was simply a bullet in his body.

"A boy who will be a great man should not think of
To overcome a thousand obstacles but to win
despite a thousand obstacles and defeats. "-
Theodore Roosevelt

4. President Nelson Mandela: He embodied the
type of leader Integrator Nelson Mandela was an
exemplary integrator. After having Spent nearly
three decades in prison for his activities
Revolutionary anti-apartheid, he became the first
South African president to be elected by elections
fully representative democracies. It has reached this
peak because of its ability to bring people together,
to connect personally and build trust by listening.
Today, there is a spider, a sea slug, a nuclear
particle bearing its name and even a day, declared
by the United Nations - July 18 Nelson Mandela
International Day, "If you want to make peace with

your enemy, work with him, make him your partner." - Nelson Mandela

5. Earnest Shackleton: Pioneer: He embodied the type of pioneer leader. An example of a pioneer, Earnest Shackleton was a polar explorer who was known to His daring accomplishments. He was designated as the "life and soul" of the ships he sailed, raising the spirits of the crew by his antics and his spontaneous, optimistic and indefatigable spirit. When selecting 26 crew members from 5,000 endurance expedition candidates, Shackleton tested singing ability in addition to more practical skills. This proved to be useful whenShip was iced and finally crushed and lost - every night.
To sing were a means of the crew maintained their Morale while living on the trapped vessel, then Polar ice for several months.
"It is in our nature to explore, to reach the unknown.

The only real failure would be to not explore at all. - Ernest Shackleton.

What type of brand will you do?

Make your style of influence flexible to stimulate your impact: You and I do not know each other. And yet, I am sure that we have something important in common. And in addition, what we have in common, we also share with leaders of all kinds, politicians, sales representatives, and my 10 year old son. What can that be, you ask? We all spend a lot of time and energy trying to influence others.

What we may not share are the strategies we most commonly use in our attempts of influence. I tend to support my point of view with evidence and data. My son, on the other hand, perfected the strategy of bringing people down through relentless demands.

Depending on your type of business chemistry, some influence strategies may be more natural to you and a little more of a stretch. While there is some power to focus on your strengths, there is also evidence that when it comes to influence, using more strategies is better, so it probably pays to work on the " Adding some of the stretching strategies to your arsenal.

1. Gain confidence by exposing your expertise:

People are often influenced by experts. But when you use expertise as an influence strategy, it is not enough just to be an expert, you should expose your expertise, share your thoughts as well as evidence that you know what you are talking about. Such evidence may take the form of references to earlier experiences or more formal references.
As an illustration of the power of expertise, a research study showed that patients in a

rehabilitation facility were more likely to follow the recommended exercises when their therapist's awards and certifications were publicly displayed. This strategy can be an advantage for Integrators for several reasons. First, the trend of integrators with nonlinear thinking and large image can meaning that they are less likely to deeply deepen a particular issue to develop real expertise. And even when they have expertise, the consensus approach of integrators - which sometimes leads them to listen more than they speak - can harm their perception as experts.

Drivers are probably more likely to influence through expertise. With their intense curiosity and cerebral nature, drivers are particularly likely to develop in-depth expertise around specialized topics. In addition, their competitiveness and attitude to take charge of them probably make them more comfortable expressing their positions strongly and displaying their status as experts.

2. Connection by displaying emotional engagement

Without emotion, people can question your commitment to an idea. On the other hand, showing passion for an idea can inflame passion in others, as emotions can be contagious. Research studies on "emotional contagion", where individuals attempt to spread positive emotion in a group, find that not only do these groups experience an increase in positive mood but they manifest more cooperation, fewer interpersonal conflicts and higher performance. Your excitement about an idea can get others excited.

This approach can be a challenge for the Guardians, who are often reserved and emotionally content. While the Guardians tend to think it is important to appear calm and composed, this self-control may inadvertently communicate to others that they are not personally committed to an idea.

Showing emotional engagement is probably more natural for the Pioneers, who are often expressive and energetic, especially when talking about ideas that excite them. The passion that a Pioneer shows for an idea signals that they are personally invested in it.

3. Connect by responding to the emotions of others:

While emotions can be contagious, incongruous emotions can have bad repercussions. Meet your audience where they are by adjusting your tone to match their emotional state. Unbridled enthusiasm may not be appropriate if others have concerns. On the other hand, an approach too Serious can alleviate the excitement of those in a mild mood. "Mirror" the behaviors and emotions of others can establish a relationship, because when someone reflects us, we perceive them as more similar, which can make us feel more connected.

Responding to the emotional state of their audience can be challenging for drivers, who are less likely than other types to be listening to the emotions of others. As a result, pilots can go ahead without detecting if anyone is with them.

It is probably easier for Integrators, who are usually more empathic, both sense and respond to the emotions of others. These are important skills for building consensus, which is the specialty of the Integrator.

4. Communicating new ideas: Getting into the unknown can be scary for some people, and ideas that are wildly original may seem too exaggerated to be taken seriously. To help facilitate change for people,

CHAPTER 9

THE CHARACTERISTICS OF A LEADER THAT SUCCESSFUL

Skillful skills:

The following leadership characteristics can be learned during the career. These characteristics are very important for leaders to succeed. This is evidence that has been demonstrated over time. Although this list is not exhaustive, it represents the most important traits for a leader in achieving success.

It is not enough simply to demonstrate that the leader embodies these characteristics in order to succeed, he must actively demonstrate these characteristics by becoming exemplary for everyone, affirming these good principles as leader. By studying these leadership characteristics Honestly your own evaluation and name these traits

That you possess and prepare yourself to embody and develop those that you lack.

 The great advantage of doing this will save you from falling into the holes that bad leaders know. Thus, you will improve for the better your leadership characteristics.

1. To become a good leader : It is necessary to work with determination and learn from its mistakes. What are your basic leadership characteristics?

2. Communication Skills: A good leader must have a high level of communication skills. He must be both a good interlocutor and an auditor. He must be able to communicate at all levels of his organization; Be able to converse with subjects and also listen to their ideas, ask and answer questions to make sure they understood what was communicated. It should also be able to gain the trust of the people who share the ideas and its important visions, allowing them to associate with its leadership

3. Interpersonal skills: A leader must have good skills Interpersonal and win the trust of his disciples. He must listen to the grievances of his supporters and give constructive feedback to their needs. The dynamics of the team must be balanced so that everyone goes in the same direction. The leader who has gained the trust and respect of his followers can use this confidence to advance the organization towards the achievements of its objectives. This leader is able to use his interpersonal skills to even improve the most difficult relationships, and keep peace in the organization.

4. Flexibility: The good leader does not sit on his laurels. It is his responsibility to challenge the status quo and push the boundaries. A good chef is flexible; It adapts to its environment; He is willing to change direction, and tactics to do things. This also extends to his leadership style - adjusting slightly to get the best of each individual.

5. Decision-making: A good leader must know how to make decisions in seconds and make timely and informed decisions at the right time. Speed in decision making is supported by the fact that they are constantly in the hand with the right real-time information, which is obtained at the source - in the face of coal. The decisions are quick but informed and the Leader actively manages while walking, without being stuck behind a desk that crosses the spreadsheets hour by hour.

6.Variety of values: They must also understand that diversity is good for businesses and more diverse, the more opportunities there will be for innovation and improvement. Leaders actively seek a mix of diversity to enhance the potential for excellence and innovation of the team / organization. With this value, everyone is treated fairly and there is no favoritism.

7. Business Acumen: Driving is not enough. The great leaders understand the needs of companies and what they mean at a deeper level, in terms of performance and how to get the most out of the situation. An important aspect here is that a good leader makes use of informal networks throughout the company, in order to get the best return for the team / organization. They understand what is needed and how to do the job. They also know how to break down the barriers to change when they arise. The leader is always a few steps ahead, understanding the issues and working on plans

8. Solver / Innovation Issue: If there are

problems and problems that prevent the tasks performed, the Leader has good foresight, sounding and root causes, not content with the status quo. He / she takes no exception to the plan as a case of "The "And" That's how it is. "The Leader seeks to understand the problem, correct it and then aspire to it. It does not happen again. The challenges are overcome with innovative solutions.

9. Outcome-oriented: It is not enough to lead, and expect the team to deliver. The Leader actively strives for perfection and focuses on delivering team performance by leading and understanding everything that happens. He / she can effectively delegate, but always set the example, raise the bar and set expectations for others to follow. Since they understand the deeper needs of the organization and how to get there, they direct and communicate this plan and vision with their teams. No stone is left to chance.

10. Confidence: One of the most important characteristics of being a good leader is trust. A good leader must be confident in everything he / she does. The Leader should not hesitate to make decisions that can be popular or unpopular. A confident leader is aware of his shortcomings and still retains his Calm, even in an emergency.

11. Practice, Practice, Practice : Repetition is the mother of the skill - Meaning, the more you practice, the better you become. The art of being a good leader is to understand what you need to improve, and work to address these areas. Great leaders make mistakes. The important thing is to learn from these mistakes and build on the leadership characteristics. Practice these characteristics every day. Make mistakes, but avoid making the same mistakes again.

I recommend you keep a daily newspaper with you.

Note on a daily basis, the problems you encountered in your work, the mistakes you made, but more importantly, what you learned. Continue to think about ways to improve your leadership characteristics and note the course, lessons learned and things you have done to improve them.

Keep recording your actions and over time you will see historical evidence of your development as a leader.

"A leader is better when people barely know that there is, when his work is done, his goal accomplished." They will then say, "We did it ourselves." Lao Tseu

CHAPTER 10

HOW SOMEONE DEMARKS

What does make some people stand out from the ordinary crowd, appearing as great leaders? Why do these people live a wonderful life, while the rest live on a day-to-day basis? The average person lives history; the great leaders forge it. However, greatness is simply a set of different attitudes and habits. You too can become great if you adopt these habits. Here is what great leaders do differently and how you can begin to apply these habits in your own life.

1. **A Vision of their Future:** Exceptional leaders are the captains of their own boat called life. They know that the boat follows their instructions and they take responsibility to give these directions.

They are the ones who shape the future by having a clear vision and take 100% responsibility for everything that happens to them ... *A man without vision is like a boat without a destination*. It sails adrift in the middle of the ocean, at the mercy of tides and waves. All great leaders have a vision and they pursue this vision with immense passion. They know exactly what they want, so they are able to get others to follow them towards their desired outcome.

2. They remain Faithful to themselves no matter what happens. Exceptional leaders follow their own inner voice when confronted with a decision. They know what is best for them and they will do whatever they think right, even in adversity. *They say their truth and they act according to what they believe to be true, even with the risk that others will incriminate them*. Exceptional leaders are authentic and congruent. That's how they gain the trust of others so easily. They are not afraid to expose themselves as they are - with their strengths as well as their weaknesses. They admit that they are human and can make mistakes. They cherish their imperfection and use it as an asset. Above all, they care about their individuality and are not afraid to show it, even to those who disagree. Exceptional leaders remain true to themselves, even if others demand compliance. They know that they are the only worthwhile person to be pleased with. They have a very strong internal validation system that guides them, so they do not need the approval of others. **"Before you become a leader, success is about succeeding in raising you up."** *When you become a leader, success is about succeeding in raising others. "Jack Weald*

3. They persevere in the face of obstacles.

Of exceptional leaders is their ability to slip into

setbacks and rejections. Many exceptional leaders have faced rejections before they succeeded in getting their ideas accepted. However, they persevered and succeeded They see obstacles as challenges and opportunities for growth, not as invitations to give up, but instead of stopping them, the obstacles have the opposite effect: they are even more determined to succeed and prove that they are right and others are wrong. Great leaders do not focus on problems and discards. Instead, they focus on solutions, They can learn and do better next time. They do not take the setbacks personally. They know that they are right - their internal validation system tells them - and they do everything necessary to convince the world of this fact.

4. They act with courage in spite of fear.
Exceptional leaders are admired for their courage. *Many people who have shown great courage have*

remained in history as heroes. But what made these people different was not their absence of fear. On the contrary . They were afraid like any other human being. **What distinguishes them is their ability to feel this fear and act in spite of it.** Exceptional people have the same fears, the same doubts, the same inner conflicts and the same emotions as everybody else. But they have learned to follow their vision, no matter how they feel. They know that they are taking action for a greater cause and this vision drives them to continue even in the face of fear. It is not that they ignore their fear.

5. **Interpersonal skills:** A leader must have good interpersonal skills and earn the trust of his followers. He must listen to the grievances of his supporters and give constructive feedback to their needs. The dynamics of the team must be balanced so that everyone goes in the same direction. The leader who has gained the trust and respect of his followers can use this confidence to advance the

organization towards the achievements of its objectives. This leader is able to use his interpersonal skills to even improve the most difficult relationships, and keep peace in the organization.

6. Flexibility: The good leader does not sit on his laurels. It is his responsibility to challenge the status quo and push the boundaries. A good chef is flexible; he adapts to its environment; He is willing to change direction, and tactics to do things. This also extends to his leadership style - adjusting slightly to get the best of each individual.

7. Decision-making: A good leader must know how to make decisions in seconds and make timely and informed decisions at the right time. Speed in decision making is supported by the fact that they are constantly in the hand with the right real-time information, which is obtained at the source - in the face of coal. The decisions are quick but informed and the Leader actively manages while walking, without being stuck behind a desk that crosses the spreadsheets hour by hour.

8. Diversity of values: They must also understand that diversity is good for businesses and more diverse, the more opportunities there will be for innovation and improvement. Leaders actively seek a mix of diversity to enhance the team's / organization's potential for excellence and innovation. With this value, everyone is treated fairly and there is no favoritism.

9. Business Outcome: Drive is not enough. Great leaders understand the needs of companies and what it means at a deeper level, in terms of performance and how to make the most of the situation. An important aspect here is that a good leader makes use of informal networks throughout the company, in order to get the best return for the team / organization. They understand what is needed and how to do the job. They also know how to break down the barriers to change when they arise. The leader is always a few steps ahead, understanding the issues and working on plans to address them before they happen.

10. Solver / Innovation Issue: If there are problems and problems that prevent the tasks performed, the Leader has good foresight, sounding and root causes, not content with the status quo. He / she takes no exception to the plan as a case of "Things just happen" and "That's how it is." The Leader seeks to understand the problem, correct it and then aspire to it. Does not happen again. The challenges are overcome with innovative solutions.

11. Outcome-oriented: It is not enough to lead, and expect the team to deliver. The Leader actively strives for perfection and focuses on delivering team performance by leading and understanding everything that happens. He / she can effectively delegate, but always set the example, raise the bar and set expectations for others to follow. Since they understand the deeper needs of the organization and how to get there, they direct and communicate this plan and vision with their teams. No stone is left to chance.

12. Confidence: One of the most important characteristics of being a good leader is trust. A good leader must be confident in everything he / she does. The Leader should not hesitate to make decisions that can be popular or unpopular. A confident leader is aware of its shortcomings and still keeps his calm even in an emergency.

13. Practice, Practice, Practice

Repetition is the mother of the skill - Meaning, the more you practice, the better you become. The art of being a good leader is to understand what you need to improve, and work to address these areas. Great leaders make mistakes. The important thing is to learn from these mistakes and build on the leadership characteristics. Practice these characteristics every day.

Make mistakes, but avoid making the same mistakes again. I recommend you keep a daily newspaper with you. Write down on a daily basis the problems you encountered in your work, the mistakes you made, but more importantly, what you learned.

14. They anticipate obstacles and find solutions: Great

leaders have a plan. They are not content merely to rush headlong, without preparation. They carve a path to their goal. In addition, they try to predict what may happen on their way, so that they can be prepared for any situation. But they do not think about all the things that can go wrong, and find ways to counter them. That way consumes too much energy and time. Besides, we can think of millions of reasons why things could go wrong, but that is not the goal. *Exceptional leaders have learned to use common sense and anticipate challenges.* They do this by observing how things work and relate to each other. They have a realistic vision and avoid overestimating or underestimating their current situation. They are not too excited, nor do they become paranoid. They succeed in looking at circumstances, situations and people and seeing them as they are. Their ability to think clearly and not be constrained by beliefs allows them to accurately anticipate barriers and find solutions in advance.

15. They spend time on what matters most : Exceptional leaders are very effective. And they have exactly the same 24 hours a day that everyone has. The difference lies in their ability to manage their time. *Great leaders spend more time on activities that are important to them and that bring them the greatest accomplishment.* Since they have a vision and a plan, they know exactly what to do to make it a reality. So they invest their energy to make things happen and create a meaningful life. On the contrary, the average person spends their time on activities that distract their attention and do not bring them long-term gains. They come for instant gratification and pleasure as much as possible. Exceptional leaders will often sacrifice pleasure in the short term for a long-term gain, as they know that this is where true happiness comes from. They have learned to delay their satisfaction, while keeping an eye on their ultimate goal, and

Take the important steps that bring them closer to their dreams. "Being a leader is about nurturing and strengthening." Tom Peters

16. They are constantly improving :
Exceptional leaders are not content with what they have. They seek to develop constantly, they continually seek to learn new skills and develop their abilities. *Great leaders are perpetual students and they never get tired of learning.* They never stop dreaming either, and set their own goals. *They have a permanent vision of how their ideal of life looks and they constantly update that image as soon as they are close to reaching it.*

17. Exceptional leaders set very high standards for themselves. Whenever they are close to achieving their goals, they set new ones, so they can continue to go farther and farther away.

They are expanding and growing, and are constantly looking for new challenges and new ways to get out of their comfort zone.

Unlike the average people who settle in comfort, exceptional leaders embrace the challenges because they know that these are the prerequisites for growth and lasting satisfaction.

"Growing and raising people is the highest leadership vocation." Harvey Firestone

CHAPTER 11

THE FIVE COUNCILS OF JULES CÉSAR

Without doubt, Julius Caesar is by far one of the most famous historical figures. And for good reason, his journey was so exceptional that he changed the foundations of the greatest power of the time to make it shine as never before.

Besides his great military conquests and his major role in the tilting of the Roman republic towards the Empire, Caesar also succeeded in attracting an ever growing affection for the Roman people. And because of his many reforms he allowed the "plebs" to regain a new dignity where the senatorial elite had previously done a lot of damage. Undoubtedly, Julius Caesar is one of the greatest leaders in history. That is why, today, we will take a closer look at this characteristic of the personality. So, to better inspire us, let's discover together the 5 elements that make Julius Caesar a true leader:

Like Julius Caesar: Know how to challenge the Statue-Quo. To succeed in creating such a change in the foundations of the Roman republic, one can easily imagine how much our dear Jules is ignoring the Statue Quo. Indeed, convinced by the soundness of the reforms he wished to undertake, he at no time lacked the courage to initiate many changes. Antagonistic behavior to the majority of the senators who, like Cicero, dared not push anything, preferring always to side with the majority.

Changing the Statue-Quo, having the courage to showcase your innovative ideas will require you to take risks and especially commit yourself. However, if you remain sure of your convictions, there is no doubt that some, sharing your ideas, will de facto rank side by side with the courageous leader you are.

1. Like Julius Caesar: Assume Your Responsibilities

It is easier to blame others and not take responsibility. Such a Caesar, a true leader, a true leader, must of itself take the burden of the decisions taken. If the result is disappointing, if the failure is present, it is up to you to assume the responsibility and not to your colleagues.

Act in this way and you will come out grown and admired by your peers.

2. Like Julius Caesar: Show the Example

Of noble class, proconsul and chief of many legions, we might imagine that Caesar had a special treatment during his campaigns. Yet, despite his status, he was the first to set an example and take part in the various difficulties an army might encounter. Redouble courage and perseverance during the battle, leave the guest rooms to the wounded and share with his comrades the makeshift lodgings, never be reluctant to work and show more courage than his own soldiers ... here are a few Examples of our general's conduct.

These various states of service have thus gained a strong adhesion of these troops, in particular the XIIIth legion which, under his command, braved the prohibitions established by the authority of the Senate and marched on Rome.

A leader is not above all ungrateful tasks! *Always show your peers the good example and they will respect you sincerely.*

3. Like Julius Caesar: behave like a leader, whatever the occasion. There is no need to be a consul to exert his influence on the Senate. And that, Julius Caesar, as his eternal rival Pompey, have both understood. We naturally tend to think that it is the job that makes the responsibilities when it is just the opposite.

Do you want to become a manager? So, behave yourself as one, and one day there will be a good chance that you will be the first person to be thought of for this position. Moreover, it is important to recognize that social status / professional status, although this is contributed to you, it is not the essential factor in determining the leader of a group. Those who seem to hold the reins are not the true leaders; they often hide behind "the counselor" or "the man of the shade".

4. Like Julius Caesar: Follow your own way and never turn away from it. A destiny that seems inaccessible is ambitious. And yet Julius Caesar never ceased to believe in his dreams and to follow his own voice, notwithstanding the most powerful senators. It takes a lot of courage to tirelessly follow his own path and never allow obstacles to break your dreams.

5. "Breakers of dreams" is your destiny and it belongs only to you! Do not let anybody hinder this road; assert with conviction your true desires and you will have everything needed for a true leader!

Hope that these 5 tips inspired by the Emperor will help you to awaken the leader who is in you.

CHAPTER 12.

10 DIFFERENCES BETWEEN A BOSS AND A LEADER

In my research I discovered this document entitled "10 Differences Between a boss and a leader" and some of their similarities. Though I believe there are other elements that distinguish the leader from the boss, I have differentiate them based on specific characteristics. I will formulate this according to my own terms and make few comments according to my way to conceive things.

1- The boss pushes his employees, the leader motivates them. While the boss orders, the leader coaches his staff based on their strengths and weaknesses, in order that each member will give the best of himself in the quest for their common goals.

2- The boss relies on authority, the leader counts on volonté. Au good place to take a stick to look after his employees, the leader conveys the values that will allow healthy and full cooperation of all.

3- The boss inspires fear, the leader inspires enthusiasm. If you are afraid of your superior, he is a boss, but if he is for you a source of inspiration, then he is a leader. Contrary to the patron who is feared, people smile in his presence and vomit as soon as he turns his back; The leader arouses enthusiasm among his collaborators.

4- The boss says "I", the leader says "we" The leader works for the collective and makes others work for the team. He does not satisfy followers, but forges other leaders around him.

5- The boss says "just in time", the leader reaches in advance. Le boss allows himself to wait, but the leader shows he is most interested in the subject of his organization.

6- Blame are from the boss, the fault is for the leader. The leader seeks to identify the fault in order to rehabilitate (s) failed (s).

7- The boss knows how to do, the leader shows how to do: The boss is limited in his academic knowledge, while the leader is the sum of experience and intelligence capable of innovating in the face of all eventualities.

8- The boss makes work drudgery, a leader makes it a passion. A leader can feel tired, but we can never feel bored. Contrary to the boss, he transmits this passion of life and work to the team.

9- The boss says "Go", the leader says "Let's go." While the boss points to the way, the leader paces the paths at the head of the pack. He communicates to his team the determination to go as far as possible.

10- The boss is justified and rejects responsibility, the leader assumes it. Instead of exonerating himself against the poor performance of his staff or one of his collaborators, the leader assumes and draws all the consequences. We can find other elements distinguishing the leader from the boss, because this list is far from being exhaustive. However it is crucial to retain the essential.

Listen what Isou maila Sikirou (Coach contractors leaders), says: "ANYONE AS A HEAD OF THE GROUP IS A PATRON. BUT ALL PATRONS ARE NOT LEADERS. THE BIGGEST DIFFERENCE BETWEEN A BOSS AND A LEADER IS THE BOSS JUST

OBSERVES HIS SUBJECTS BECAUSE OF HIS]AUTHORITY, BUT THE LEADER IS RESPECTED AND ADMIRED AS A MODEL, NOT ONLY BECAUSE OF ITS EXPERTISE BUT MAINLY IN REASON OF INTRINSIC QUALITIES (CHARACTERS, ATTITUDES, INTELLIGENCE, EMOTIONAL ...). THOSE WHO ASPIRE TO BECOME LEADERS MUST FOLLOW THIS EXAMPLE ... "

Let all of us become leaders, and abandon the position of boss. That way our organizations will be on their way to the rendezvous of the accomplished.

CHAPTER 13

WHAT TO LEARN TO BECOME A BETTER LEADER

Anyone and even a fictional character is able to provide valuable ideas that can help you develop the characters of a good leader.

This person can help you to shape within you some important characteristics like tenacity, honesty and ingenuity among others. These people were not my mentors by design; But without them, I would have missed the lessons of tenacity, honesty and ingenuity. Here are the five key points to learn from a mere mentor:

1. Cultivate toughness: There is a link between physical work and mental work. Splitting wood or clearing a corner of the bush requires as much tenacity, hard work and discipline to complete this physical task as to face up to mental challenges, such as reflections to make a great project come to fruition.

Remember people in your life who have overcome great physical challenges. They did not give up their first commitments; But worked hard and pressed to the end. Learning this tenacity will help you tame on failures throughout your life.

2. Cultivate friendships based on honesty.

The concept of friendship has changed a lot lately, but when you think about your true friends, it can only be those with whom your friendship is based on honesty. It is only with such people that you share authentic links.

Remember that your friends are influencing your life. Concentrate on friends who make positive contributions especially when you encounter obstacles in life. Good friends always give good advice.

3. Learn to solve your problems: As a leader, you face challenging challenges. Any challenge faced finds a solution. If you work hard to overcome them, then there is always a lesson to learn.

4. Become fully committed: One must devote the best of either even on any subject that ultimately will bear your name. Remember that what you do represents who you are and you are what you spend the time that God gives you on this earth. An example, when the Italian painter Michelangelo was painting the ceiling of the Sistine Chapel in the 16th century, a visitor watched him as he climbed to work on a corner of the painting, went down to watch and went up and down . The visitor who was impressed by the whole painting asked him the question of knowing the great painter was bored so much for a small error that no visitor could unseal. Michelangelo replied, "If you do not see this error, I see it and will not stop working on this corner of the painting only when I am totally satisfied. Every best leader must have Michelangelo's heart and eye in everything he does. He must from time to time stop and look back upon all that he has already accomplished and be his best judge; Review these methods to understand how to improve.

5. Absorb knowledge through books: A better leader is one who continues to learn from others. Reading books and learning and listening to others, and adapting new methods because of the social and technological evolution puts you on the path to success.

6. Leadership must be focused: To be a good leader, you must spend a lot of time concentrating on the major topics and being less distracted. Try to selectively ignore some criticisms and see if they do not touch the major orientations of the organization.

7. Trust: "A leader inspires confidence and" follow-up "by having a clear vision, showing empathy and being a strong coach. As a female leader, to be recognized, I feel that I must be arrogant and assertive, but always try to maintain my southern education, which emphasizes kindness and generosity. The two work well together to earn respect. "

8. Be transparency: "I never redeemed the concept of" wearing the mask ". As a leader, the only way to create trust and buy-in from my team and colleagues is to be 100% authentically open, Sometimes flawed, but always passionate about our work.

It allowed me the freedom to be fully present and

coherent. They know what they get at all times. No surprises. "- Keri Potts, Senior Director of Public Relations, ESPN

9. Integrity: "Our employees are a direct reflection of the values we embody as leaders. If we play from a reactive and obsolete playbook need to be right instead of doing what is right, then we limit the full potential of our business and

10. Communication: "If people are not aware of your expectations, and they are short, it's really your fault for not expressing it to them. You must be in constant communication with the people you work with . But communication is an act of balance - you may have a specific need to; it is important to treat the workplace as a collaboration – the leader always wants people to tell him their thoughts and ideas - that's why he all has these very talented people who work with him.

11. Accountability: "It is much easier to blame than to hold you accountable. But if you want to know how to do it well, learn from financial expert Larry Robbins who wrote a truly humble letter to his investors about his poor judgment that caused their investments to falter. He then opened a new fund without management fees and performance . It is character. It is accountability. It is not just taking responsibility; It is the next step to do it right.

12. Concern : "It takes real leadership to find the strengths within every person on his team and then be prepared to look outside to fill the gaps. It is best to believe that your team alone does not have all the answers - because if you believe this, it usually means that you are not asking all the right questions.

Leadership is one of those nebulous terms - you hear it all the time, but it has various definitions. The traits that make up a good leader can vary depending on the organization, team, manager and work environment.

CHAPTER 14

NINE TREATS TO DEFINE A GREAT LEADER

To motivate your team to achieve the highest levels of performance, and create an extraordinary organization in the process, are what require qualities one needs to model each day.

Many leaders are competent, but few qualify as remarkable. If you want to join the rank of the best of the best, make sure you embody all these qualities all the time. It's not easy, but the rewards can be truly phenomenal.

1. Awareness: There is a difference between management and employees, employers and workers. Leaders understand the nature of this difference and accept it; that informs their image, their actions and their communication with others. They behave in a way that distinguishes them from their employees - not in a way that suggests that they are better than others, but in a way that allows them to maintain an objective perspective on everything that happens in their organization.

2. Discernment: All leaders must make tough decisions. This goes with the work. They understand that in some situations, difficult and timely decisions must be made in the interests of the entire organization, decisions that require firmness, authority and purpose that will not appeal to everyone. Extraordinary leaders do not hesitate in such situations. They also know when not to act unilaterally, but rather to encourage collaborative decision-making.

3. Empathy: Extraordinary leaders praise in public

and deal with problems privately, with real concern. The best leaders guide employees through the challenges, always looking for solutions to promote the long-term success of the organization. Rather than doing personal things when they encounter problems, or blaming individuals, leaders look for constructive solutions and focus on advancement.

4. Liability : Extraordinary leaders take responsibility for the performance, including their own. They monitor all outstanding issues, audit employees and monitor the effectiveness of the company's policies and procedures. When things are going well, they praise. When problems arise, they identify them quickly, seek solutions and put things back on track.

5. Trust : Not only are the top leaders confident, but their confidence is contagious. Employees are naturally attracted to them, seek their advice and feel more confident as a result. When challenged, they do not give up too easily because they know that their ideas, opinions and strategies are well informed and the result of a lot of work. But when proven false, they take responsibility and act quickly to improve situations under their authority.

6. Optimism : The best leaders are a source of positive energy. They communicate easily. They are intrinsically useful and genuinely concerned about the well-being of others. They always seem to have a solution, and always know what to say to inspire and reassure. They avoid personal criticism and pessimistic thinking, and seek ways to reach consensus and get people working together effectively and efficiently as a team.

7. Honesty: Strong leaders treat people the way they want to be treated. *They are extremely ethical and believe that honesty, effort and* *reliability are the foundation of success.* They embody these values so openly that no employee doubts their integrity for a minute. They share information openly, and avoid rotation control.

8. Concentration : Extraordinary leaders plan ahead, and they are supremely organized. They think about multiple scenarios and possible impacts of their decisions, while considering viable alternatives and making plans and strategies - all targeted to success. Once prepared, they establish strategies, processes, and routines for high performance to be tangible, easily defined, and monitored. They communicate their plans to key players and have contingency plans in case last-minute changes require a new direction (which they often do).

9. Inspiration : Put it all together, and what

emerges is an image of the truly inspiring leader: someone who communicates clearly, concisely, and often, and by making it motivates everyone to give their best all the time. They challenge their people by setting high but achievable standards and expectations and then giving them the support, tools, training and latitude to achieve those goals and become the best employees they can be.

CHAPTER 15

TEN IMPRESSIVE QUALITIES OF GREAT LEADERS

Great leaders come in all shapes and sizes, and they are certainly not limited to any particular industry or Fortune companies. A great leader can be anyone from a politician to the owner of a small cafe. However, despite the enormous differences in professions and backgrounds, there are some things that all great leaders have in common. The question is, do you see one of these qualities in yourself?

1. Positive Attitude : Great leaders know that they will not have a happy and motivated team unless they show themselves a positive attitude. This can be done by staying positive when things go wrong and creating a relaxed and happy atmosphere in the workplace. Even

simple things like providing cupcakes or Beers on Friday can make the world of difference. An added benefit is that team members are likely to work harder and do overtime if necessary if they are happy and appreciated.

2. Managing failures and successes : No matter how we try to avoid them, failures will occur. We just need to know how to deal with them. The great leaders follow them. They remain calm and logically think of the situation. They use their resources. *What they do not do is fall apart and reveal to their team their anxiety.* This leads to total demoralization of the team. Great leaders actually lead, even when they are faced with setbacks such as when they confront successes.

3. Takes Responsibility : Great leaders know

that when it comes to their business, the workplace or whatever the situation, they have to take personal responsibility for failure if there is one, waiting for

employees to feel themselves, If they do not do it themselves. The best leaders do not apologize; They take the blame and then find out how to solve the problem as soon as possible. This proves they are trustworthy and possess integrity.

4. Develop those around oneself: *Any good leader knows how important it is to develop the skills of those around them.* They recognize their skills from the start. Not only do they develop them to make their work easier, but they improve them better and better even in the moral sense. In addition, they can develop certain skills that employees did not have at the outset, skills that will benefit the workplace. They share their knowledge with their team. They give them the opportunity to reach new levels. The employees respect them and keep them loyal because of all this.

5. Capable to delegate : Whatever the case, no leader can do everything by himself. Even if he can, why do when surrounded by a group of others ? Good leaders recognize that the delegation does more than simply relieve their own stress levels (although this is obviously a nice bonus), but it shows others that they are confident in their abilities. This leads to a moral elevation of the place and the level of loyalty of his staff. This one wants to feel appreciated and confident. But if the leader does not delegate a few responsibilities, not only will he be too busy, his health will suffer and he will not have the true collaboration of his team.

6. Knowing how to communicate :
It's much more complex than it actually sounds. Good communication skills are essential for a great

leader. One can very well understand the crazy cave that is your brain, but that does not mean you can take the ideas that are in your memory and explain them and explain them to someone else. All employees can not be idiots if this problem repeats and people do not seem to do better in your business. The best leaders must be able to communicate clearly with the people around them. They must also be able to interpret others correctly and not believe that they are understood without proof that others have understood him. If it is an area with which he has difficulties, a good point must be established at the outset to promote communication between him and the members of the team in order to establish an open door policy.

7. Proof of Confidence : All great leaders must show an air of confidence to succeed. Self-satisfaction and arrogance should not be confused. Trust is important because people will know how to behave, especially when things do not go 100% in the direction it takes. *If you remain calm and balanced, your team members will be much quieter. As a result, morale and productivity will remain high and the problem will be resolved more quickly.* If you panic and you blame everyone, rest assured that everything will collapse. So be confident - without being an idiot.

8. Using Intuition : It is safe to say that all great leaders will have to enter unexplored waters at some point during their career (figuratively, of course). For this reason, they must be able to trust their intuition and rely on past experiences to guide them. On top of that, they have to trust their own intuition to guide their staff or team members.

9. Have a sense of humor : It is imperative for any type of leader to have a sense of humor, especially when things go wrong. The members of the team will look for their leader to know how to react in an apparently terrible situation. It would probably be better if the leader is in the corner. He must be able to laugh at things. *If staff morale falls, productivity will also fall.* You must work to establish such an environment before any kind of encouraging or personal discussions in the workplace. After all, your workplace is not a Russian gulag. At least, I really hope that is not the case.

10. Demonstrate commitment : This has a double meaning: First, one must respect one's commitments and promises. If the leader does not do that, he will find out that people will not have confidence in him. Then he must be a most committed and hardworking person in the group. He must set an example.

All the great leaders remain commit. Why should staff and team members give themselves entirely to work when the leader does not? Proving his own commitment, he will inspire others to do the same, and more than that, he will earn their respect and inculcate in them a good work ethic.

CHAPTER 16

8 CHARACTERISTICS OF GREAT LEADERS

A. Good leadership is essential to businesses, government and the many groups and organizations that shape our way of life, work and play.

B. Having an excellent idea, and assembling a team to bring this concept to life is the first step in creating a successful business.

C. Although finding a new and unique idea is quite rare, the ability to successfully execute this idea is what separates dreamers from entrepreneurs. And that's where leadership turns potential into reality.

D. Leaders are a key human resource in any organization. We usually think of competing companies through their products, but they are probably more competitive through their leaders than their products.

E. The best leaders develop better employees and the two together develop better products.

F. The importance of leadership in management can not be overstated. To do things by people, management must provide leadership in the organization.

G. Teamwork is essential to achieving organizational goals. Managers must influence the team to accomplish the job through leadership. Second, leadership helps authority.

To get an insight into the importance of leadership in business success, Mark Bilton, founder of Thought Patrol, one of the leading Australian authorities on the subject, gives us an impressive track record of his ability to take a business or a totally broken organization and turn it into success.

Below are his 8 most common characteristics of great leaders.

1.Collaboration Trust in transparency : If you are connected to your team and sincerely interested in their participation and well-being, its members will join you in your quest. *People possess what they help create.*

2. Visionary : A certain named Hoffer said **that "the leader must be practical and realistic but must speak the language of the visionary and the idealist."** Here lies the balance : Leaders must live in the future state and bear the vision while engaging in the deliberate motivation and practical realities of the present.

3. Influence : The key to successful leadership today is influence, not command and control authority.

4. Empathy : We are all imperfect human beings; we all have our bad days. *Leaders, who recognize that he or she leads full and real people, not only manage for a result, will engender an enormous amount of fidelity, commitment and productivity.* <u>Treating others, as we would like to be treated</u>, is a universal principle that has been working well for over 2000 years!

5. Innovative : *"Innovate or die"* is a truism that is probably more relevant now, then at any time since the Industrial Revolution. As Dr. Gary Hamel so rightly says; **"Innovation in management will be the most sustainable source of competitive advantage."** There will be many rewards for cutting-edge companies.

6. Earthing : Leaders need to be focused and balanced to be effective and resilient. We must be mindful of caring for ourselves so that we can effectively serve others. We are the body, soul and spirit and each leader will have to ensure their own fundamental well-being, in order to be durable in the cauldron of the modern workplace.

7. Ethics: The unpleasant choice of your moral framework resembles a odd business success driver. Taking an early position can save you, and others, a world of evil is a more sustainable life and choice of businesses. Often, **it's as simple as doing what you say you will do.**

8. Passionate : *The face stone leader who shows no emotion is a relic of the industrial era. The true passion for the people and purpose* *are great motivators that build momentum.* A committed, empowered and led team with a clear vision and goal set by a passionate leader are forces to be counted on.

These 8 qualities are those who help Mark have a so impressive track record for saving companies that perish. This is the message leaders need to learn from heart especially for our fast moving business world.

CHAPTER 17

SEVEN OTHER QUALITIES OF
A GREAT LEADER

Some of the qualities we will cover in this book and many chapters may have been covered in others. They keep coming up in order to show their importance in good and successful leadership.

1. A good leader is exemplary : It is of the utmost importance that a leader be trusted to lead others. A leader must have confidence and be known to live his life with honesty and integrity. A **good leader "walks the talk"** and in so doing gains the right to have responsibility for others. *True authority arises from respect for the* *good character and reliability of the person who leads.*

2. A good leader is enthusiastic about his work : He shows enthusiasm for his work or cause and his leadership role. People will respond more openly to a person of passion and dedication. Leaders must be able to be a source of inspiration, a motivator for the action or the required cause. Although the responsibilities and roles of a leader may be different, the leader must be considered as part of the team working towards the goal. This type of leader will not be afraid to roll their sleeves and get dirty.

3. A good leader is confident : In order to direct and define direction, the leader must appear confident as a person and in the leadership role. Such a person inspires confidence in others and attracts the trust and best efforts of the team to properly accomplish the task. A leader who imparts trust to the proposed goal inspires the best effort of the team members.

4. A leader must be ordained, determined: A leader must function in an orderly and determined manner in situations of uncertainty. People turn to the leader in times of uncertainty and lack of familiarity and find security when the leader portrays confidence and positive behavior.

5. A good leader is tolerant, calm and composed: Good leaders are tolerant of ambiguity and remain calm, composed and determined to the main objective. Storms, emotions and crises come and go and a good leader takes them as part of the journey and keeps a cool head.

6. A good leader is capable and thinks analytically: A good leader main focus is to focally be able to think analytically; to see a situation as a whole and be able to divide it into sub-parts for closer inspection. He not only has the goal in view, but breaks it down into manageable steps and makes progress to solve the whole situation.

7. A good leader is committed to excellence: A good leader is committed to excellence to lead to success. He maintains high standards and is proactive in raising the bar in order to achieve excellence in all areas.

These seven personal characteristics are fundamental

for good leadership. Certain characteristics may be more naturally present in the personality of a leader.

However, each of these characteristics can also be

developed and reinforced. A good leader, whether they naturally possess these qualities or not, will be diligent to develop and reinforce them constantly in their role of leader.

The Author: Barbara White, Leadership Expert. Current President of Beyond Better Develop, specialty: motivation and training in interpersonal relations.

CHAPTER 18

WHAT A POSITIVE LEADER DOES Its 7 Principles of Application

Small actions can have big effects in organizing a company, small or large. **THERE ARE MORE THINGS THAT CAN BE MADE TO PUT PEOPLE UNFORGETTLY, AS THAT IS THAT THEY ARE CONTRARY**; For example, being unrecognized or unrecognized may be more valuable to a certain employee than having an increase in wages.

Jane E. Dutton and Gretchen M. Spreitzer, two scientists in business organization, published a book entitled **"How to Be A Positive Leader"** in 2014, this book gives us seven principles to strengthen positive leadership to better stimulate employees.

Yes, it is possible to create a profound change in the company through simple actions to positively change the attitudes of employees.

1. Encourage quality interactions: Creating dynamic teams for competitive advantage in a company aimed at increasing salaries for example can be a source of curiosity and motivation for learning new things;

To achieve this goal, two strategies can be put in place: **First,** the leader needs to show employees that he respects and values them. This can only be done when he pays attention to what they say and remains POSITIVE when they express their opinions. **Second,** it must facilitate teamwork and positive interactions encouraging employees to play more games (team building, orienteering).

2. A sensible output to create motivation:
When an employee assumes a significant position
that has a positive outcome for others outside the
company it improves his quality of life. This fact
considerably increases his intrinsic motivation.
Allow employees to see the impact of their work,
connect them to users of their end products by
organizing a small party and inviting customers to
testify in person of their satisfactions to employees.
Ask employees from time to time to report on the
action that has contributed to the success of the
business.

3. Cultivate Positive Identifiers: POSITIVE
IDENTITY refers to all these elements that help
people feel happier and more concentrated, which
translates into better work. But then what makes
people feel good? A framework that explains this

positive identity is the model "**GIVE**", which has four elements.

A / - Growth: Man tends to feel better when he evolves, progresses and feels that he is getting closer and closer to the man he would like to be.

B / - Integration: When man makes a good part of things in his life - as in his work, his family and his passions - this promotes the development of his Positive Identity.

C) Virtuous: If he is virtuous because of his actions which naturally define his positive identity; all that gives a sense of contribution and benevolence.

D / - Esteem: When he feels that his personality is appreciated by those around him.

A positive leader can help employees improve their Positive identities by encouraging them to take advantage of their strengths and virtues in the place of work. He can do this for example by asking an employee to tell him a story in which he felt better. By doing so, the employee will have an opportunity to talk about his POSITIVE POINTS of his forces that can be HELPFUL for the leader.

If the employee finds out that he is empathetic, the leader can easily find a situation where this empathy can be expressed, making him a mediator between two employees in conflict within the company.

A positive leader should be able to constantly make good use of the good qualities of his team helping members develop their POSITIVE IDENTITY.

4. Admit that everyone brings his color:

When the company treats every employee in much the same way, no matter how grateful we are all DIFFERENT and UNIQUE, such a company must let its employees design their own work in order to get the best bet from its staff.

"Designing your work is about letting the employee adjust and personalize his role within the company so that he can respond according to his passions, values and abilities. This is changing my job rather than changing EMPLOYMENT.

This benefits not only the employees, but also the company. The results of such a spirit will be: the employees will be less absent, will work much better and will be more PROACTIVE.

How will an employee design his or her job? It is a variable concept that can work at different levels.

CHAPTER 19

THE PROFESSIONAL QUALITIES OF
A GOOD LEADER

Here's what five leadership professionals see as traits that make up a good leader: Rachael Fisher-Layne, vice president of media relations, a public relations agency

1. Honesty. Always do the honest thing. That makes employees feel like they know where they stand with you at all times.

2. Concentration. Know where you are going and have a strong feeling that seems to be a calling to bring others to a given place. If you are not sure, how can your people be safe? You must have strong focus and stay the course.

3. Passion : Anyway, you must have passion for what you do. Live, breathe, eat and sleep your mission.

4. Respect : Do not play favorites with people and the treatment of all people - regardless of the station in life, what class or what rank in the flowchart - the same.

5. Excellent ability to persuade. People must believe in you and your credibility. The image is everything and the belief that people have in you, your product, your mission, your facts or your reputation are the key to being a great leader. We have to persuade people of this - it does not happen.

[Darcy Eikenberg, a leadership and work coach, Red Cape Revolution]

6. Confidence : If you do not believe in yourself, no one will. I hear leaders worry that if they show too much confidence, others will find them arrogant. The reality is that people want to know what you know for sure – and what you do not have. Having the confidence to say **"I do not know"** is a powerful skill.

7. Clarity : The only way to get confidence is to really become, really clear about who you are and what is most important to you. New leaders fail when they try to become all things at all, or try to do too much of their field of excellence. **Clarity helps you say "yes" to good things - and "no" to others.**

8. Care : The strongest and most effective leaders I have encountered are not only interested in business, but also in the people who live there and in the people who suffer from it. In addition, they show that they care about their words and actions, even prove how they care for themselves and their family by taking vacations unplugged and pursuing their own professional development. Care should not be a four-letter word in our workplace today - and the best leaders know it.

Tom Armor, co-founder, High Return Selection, a recruitment firm.

9. Integrity : They are people who are respected and worthy of being listened to.

I find that in general, because of all the economic difficulties, employees give priority and seek leaders and organizations that are honest and respect their commitments.

10. Compassion : Too many leaders these days manage with the balance sheet, often to the detriment of their employees and long-term relationships with customers. Talented people want to work for leaders and organizations that truly care about their employees and the communities in which they operate.

11. Vision and shared actions : People produce real business gains and intelligent people need to understand what is needed and be part of the solution.

12. Commitment : Great business leaders are able to get all the members of their teams engaged. They do so by offering them challenges, seeking their ideas and contributions and making them acknowledge their contributions.

13. Celebration : In today's working environment,

people work very long hours and they need to take some time to celebrate their successes in order to recharge their batteries. Leaders who fail to make the environment become "burnout" because of overtime work.

14. Humility : True leaders have confidence, but they realize the point where they become pride.

15. Empowerment : True leaders make their associates feel emboldened and powerful, not diminished and helpless.

16. Collaboration : True leaders solicit comments and comments from those around them so that everyone feels part of the process.

17. Communicative : True leaders often share their vision or strategy with those around them.

18. Interspersed or courageous : True leaders are not afraid to take risks or make mistakes. True leaders make mistakes arising out of risk.

CHAPTER 20

THE 7 THINGS THAT GREAT LEADERS DO THE SAME WAY

"A leader is better when people barely know that he exists, when his work is done and his goal is reached, and they will say," We did it ourselves. " Lao Tseu

Ordinary people live history as the great leaders forge it. However, greatness is simply a set of different attitudes and habits. You too can become great if you adopt them. Here is what great leaders do differently and how you can begin to apply these habits in your own life.

1. They Dedicate For Their Future : Leaders

are exceptional people; They are captains of their own boat called life. They know that this boat follows their instructions and they take responsibility for giving it directions. They are the ones who shape the future by having a clear vision and take 100% responsibility for everything that happens to them ... A man without vision is like a boat without a destination. It sails adrift in the middle of the ocean, and is at the mercy of tides and waves. All great leaders have a vision and they pursue this vision with immense passion. They know exactly what they want, so are able to get others to follow them towards its desired outcome.

2. They remain faithful to themselves no matter what happens : Exceptional leaders follow their own inner voice when confronted with a decision. They know what is best for them and they do everything they think right, even in adversity.

They say their truth and they act according to what they believe to be true, even with the risk that others will incriminate them. Exceptional leaders are authentic and congruent. That's how they gain the trust of others so easily. They are not afraid to expose themselves as they are - with their strengths as well as their weaknesses.

They admit that they are human and can make mistakes. They cherish their imperfection and use it as an asset. Above all, they care about their individuality and are not afraid to show it, even to those who disagree.

Exceptional leaders remain true to themselves, even if others demand compliance. They know that they are the only worthwhile person to be pleased with. They have a very strong internal validation system that guides them, so they do not need the approval of others. When you become a leader, success is about succeeding in raising others.

3. They persevere in the face of obstacles :

One of the most important traits of exceptional leaders is their ability to slip on setbacks and rejections. Many outstanding leaders were confronted with rejections before they were able to get their ideas accepted. Yet, they persevered and succeeded.

What brought them to success is their state of mind. They see obstacles as challenges and opportunities for growth, not as invitations to give up. Instead of stopping them, the obstacles have the opposite effect: they are even more determined to succeed and to prove that they are right and the others are wrong. Great leaders do not focus on problems and rejections.

Instead, they focus on solutions, on what they can learn and do better next time. They do not take the setbacks personally. They know that they are right - their internal validation system tells them - and they do everything necessary to convince the world of this fact.

4. They act with courage despite fear :
Exceptional leaders are admired for their courage. Many people who have shown great courage have remained in history as heroes. But what made these people different was not their absence of fear. On the contrary . They were afraid like any other human being. What differentiates them is their ability to feel this fear and act against it.

Exceptional people have the same fears, the same doubts, the same inner conflicts and the same emotions as everybody else. But they have learned to follow their vision, no matter how they feel. They know that they are taking action for a greater cause and this vision drives them to continue even in the face of fear. It is not that they ignore their fear; In fact, they recognize it – because they recognize their weaknesses and are comfortable with exposing their vulnerability - but they do everything that is most important to them and they do not allow fear to paralyze their actions. They use fear as a catalyst that propels them in the desired direction.

5. They anticipate obstacles and find solutions. Great leaders have a plan. They are not just satisfied to bow his head down, without preparing himself. They carve a path to their goal. In addition, they try to predict what may happen on their way, so that they can be prepared for any situation.

But they do not think about all the things that can go wrong, and find ways to counter them. This consumes too much energy and time. Besides, we can think of millions of reasons why things could go wrong, but that is not the goal.

Exceptional leaders have learned to use common sense and anticipate challenges. They do this by observing how things work and relate to each other. They have a realistic vision and avoid overestimating or underestimating their current situation. They are not too excited, nor do they become paranoid. They succeed in looking at circumstances, situations and people and seeing them as they are.
Their ability to think clearly and not be constrained by beliefs allows them to accurately anticipate barriers and find solutions in advance.

6. They spend time on what matters most :
Exceptional leaders are very effective. And they have exactly the same 24 hours a day that everyone has. The difference lies in their ability to manage their time. Great leaders spend more time on activities that are important to them and that bring them the greatest accomplishment. Since they have a vision and a plan, they know exactly what to do to

make it a reality. So they invest their energy to make things happen and create a meaningful life. On the contrary, the average of people spend their time in activities that distract their attention and do not bring them long-term gains. They come for instant gratification and pleasure as much as possible.

Exceptional leaders often sacrifice pleasure in the short term for a long-term gain, as they know that this is where true happiness comes from. They have learned to delay their satisfaction, while keeping an eye on their final goal, and take the important steps that bring them closer to their dreams.

Quote: "Being a leader is about nurturing and strengthening." Tom Peters

7. They are constantly improving : Exceptional leaders are not content with what they have. They seek to develop constantly, they continually seek to learn new skills and develop their abilities. Great leaders are perpetual students and they never get tired of learning.

They never stop dreaming either, and set their own goals. They have a permanent vision of how their ideal of life looks and they constantly update that image as soon as they are close to reaching it. Exceptional leaders set very high standards for themselves. Whenever they are close to achieving their goals, they set new ones, so they can continue to go farther and farther away. They are expanding and growing, and are constantly seeking

New challenges and new ways of getting out of their comfort zone.

Unlike average people who settle in comfort, exceptional leaders embrace challenges because they know that these are the prerequisites for growth and lasting satisfaction.

Quote: "Growing and raising people is the highest vocation of leadership." Harvey Firestone

CHAPTER 21

THE 5 FALSE OF A BAD LEADER

1. He is not True to himself: You must be clear about what your values are and must be consistent in their application. As part of this, you must have the courage to keep faithful to them. You must not lose sight of reality. Lost values can be one of the biggest causes of failures for bad leaders.

2. He has no consciousness in himself: You need to be clear about what your strengths are and what complementary strengths you need from others. This includes understanding others and learning the best way to use their strengths. Many unsophisticated leaders think that everyone should be like them;

That can also cause their failure. They surround

themselves with people like them. "The group thinks" can blind them and cause failure.

3. He does not know how to leverage the strengths of the team : Part of the awareness is not to expect people to change. If you think you can change someone, think again. This does not mean that you can not help them grow and develop. But do not expect to change someone (even yourself) mentally behaves. We are who we are. Your job as a leader is to understand the strengths of each person and place them in positions where they can flourish and develop. If you are good at this, you have a huge part of the equation for success.

4. He does not know how to exploit the transition period in leadership. Moving from the individual contributor to the supervisor is only the first of several transitions along the leadership pipeline. You need to understand the business model, how it applies to your current position, what you need to do to deliver the greatest value, and how to leverage your strengths at

this level. This requires building skills and focusing on the right things. No one ever tells you that there are many levels and many adjustments you need to make along the way.

5. He does not know how to build his own support. You need to foster a positive environment that allows your team to flourish. Also by aligning the reward and recognition systems that best fit your team's profile and produce results.

CHAPTER 22

THE EIGHT (8) ENEMIES FOR THE
SUCCESS OF A LEADER

Enemy 1: Manage before managing the organization: Do not behave at first sight as the one who knows everything; Not being isolated, being too aggressive, being out of self, repeating old habits and neglecting well-being.

To conquer the inner enemy, leaders must proactively manage stress behaviors in order to remove the barriers of connecting with others and build vital relationships that are essential in a new role.

Enemy 2: Do not submit to chaos : One must not remain eternally in the feeling as if all that one does is extinguished because of the answers of the daily crises.

Consideration must be given to the consequences, a leader may find it easier to submit to daily pressures than to retreat and priority to the learning needed to truly understand the key elements of the role that are well enough to fully contribute to its role. To manage chaos, a leader benefits most by taking a global approach to assessing and prioritizing learning to strategically lead him to a new role. Areas for careful consideration and prioritization are understanding of the organization, culture, business, expectations of managers, key stakeholders, peers and the team.

Enemy 3: Poor reading of crop indices :

Great leaders are enemies of the status quo. The most effective leaders see themselves as having a positive attitude. We must not denounce ourselves as the enemy of many. He has to do everything to

arrange everyone and himself against one enemy;
Watching others as enemies

of the status quo. STATUS QUO". The present state of affairs aims to preserve the status quo" Voltaire said: "Good is the enemy of the great."

"Good is the enemy of the great. Few people are reaching large lives, largely because it is simply so easy to settle for a good life.

If we want our individual organizations, our peers and our employees develop, we must realize that it depends on our willingness to fight the status quo and change. Do not change for reasons of change, but introspective consideration, thoughtful to change to win excellence.

This enemy is the one who is most likely to cause failure. Why? Because people will reject a leader who does not adapt successfully to fit, which causes a leader to disconnect in most relationships. To master the clues of culture, it is best to understand the culture of First, then to seek to carry out the desired change - with others, to provide information and get on board along the way.

Enemy 4: The Deep Problems with His Manager : The problems between the leader and the manager are very common in companies and social organizations. It is difficult to find a synchronism between these two leaders. Often it is because the leader waits that his manager spends time with him and makes things clear to him, forgetting that a passive role like that of a new leader takes time to become clear to another so Develop a productive relationship.

To calibrate with its manager a leader must assert its influence to succeed. Hold regular meetings and adapt to the manager's communication preferences. We must seek clarity in expectations and negotiate the time to learn and understand the business.

Enemy 5: Surprising Stakeholders and Peers
: Avoid giving yourself time to work with peers and key collaborators. You have to immerse yourself in the crisis of the day or be under pressure by the strategic imperatives given to leasing by a manager. Do not wait for relationships, as peers and key stakeholders have vital information about the manager of the leader, clients, culture and organization. Once connected, these people will share vital information.

Connecting with stakeholders and peers is to enable these people not only to contribute to learning the leader, but also to advocate for the success of the leader.

Enemy 6: Do not alienate your team : Never neglect to devote time to your team every time. A leader who does not take time risk Alienate and disengage team members and the potential for turnover of the best talent.

To engage their team, it is essential that team members have ties with their leader and believe their interests are best served. If this happens, the members

---

Enemy 7: Under-optimize his vision and plan. Never overlook the visions and plans that have influenced your team. Do not change vision or plan while these two are already established and are running. Sub-optimizing your vision because you have it badly articulated is your fault and you will question your responsibility and destroy the direction, leading to confusion. Therefore, the performance of the team will be reduced. In the end, the team will disengage without an inspiring picture of the future. To inspire a vision and plan, the leader must first take the time to conquer the first six enemies. It is these six first conquests that will allow him to consider and clarify the future to be created for his team. With the vision and plan, the leader will identify the high impact priorities that guarantee success to be considered by the team as strategic operational. If the vision is convincing and well articulated, and that it incorporates the comments of those mentioned above, the leader can fully engage the team in order to move towards high levels

Enemy 8. Do not celebrate the positive results : Landing is the celebration of the promotion sought: to be sure, but what happens after that is what makes or breaks these wonderful opportunities. By understanding and conquering the seven enemies of success, a leader has every chance to succeed.

CHAPTER 23

SEVEN (7) CHARACTERISTICS OF THE LEADERSHIP

Studying the characteristic of leadership is useful because we tend to break things into features to make the big concepts easier to handle. There are common traits that define leadership, and finding them takes only a few studies of those that have been successful. By activating on these traits, you can become a stronger leader. Here are some of the most common features in leadership characteristic:

1. Empathy : *Creating a legitimate relationship with your staff makes it less likely that personal problems and resentment can infiltrate and derail the group.* When your team knows that you are empathic with their concerns,

they will be more likely to work with you and share

your vision, rather than fostering negative feelings.

2. Consistency: Being a consistent leader will earn

you respect and credibility, which is essential to

getting the group's membership. By giving the

example of fairness and credibility, the team wants

to act in the same way.

3. Honesty : Another characteristic of leadership

that lends itself to credibility. Those who are honest,

especially about the concerns, make it much more

likely that barriers will be addressed rather than

avoided. Honesty also allows for better evaluation

and growth.

4. Direction, Direction : Having the vision to go

beyond the norm and aim for great things - then the

means to set the necessary steps to achieve it - is an

essential characteristic of good leadership. By seeing what can be and managing goals on how to get there, a good leader can create impressive changes.

5. Communication : Effective communication helps to keep the team working on the right projects with the right attitude. If you communicate effectively on expectations, questions and advice, your staff will be more likely to react and achieve your goals.

6. Flexibility : Not all problems require the same solution. By being flexible to new ideas and open enough to consider them, you increase the likelihood that you will find the best possible answer. You will set a good example for your team and reward good ideas.

7. Conviction : A strong vision and willingness to

see it is one of the most important characteristics of leadership. The leader who believes in the mission and works towards it will be a source of inspiration and a resource for their followers. Of course, there are several other theories on leadership and leadership styles where different skills come into play. But whatever your approach, if you view the previous traits, you will be well equipped to lead a successful team. If you would like to learn more about leadership, subscribe to our newsletter! Learn what main styles match your strengths and leadership characteristics better here.

THE SIX DANGERS THAT MAINTAIN THE

LEADERSHIP

Leaders are key elements in organizations because of their ability to maintain a united and motivated team. The key to success is to detect the worst enemies of leadership that will lead them to mediocrity.

Here are the six dangers that can threaten your leadership:

1. Hesitation : The leader must be convincing when making decisions. It is not about acting as if you know everything, but it is essential to get out of the doubts and not hesitate before the members of your team. Hesitating to give an order or to make a decision will discourage your team because it destroys the image of firmness and certainty that the leader must radiate.

2. Uncertainty : Indecision is one of the most serious threats to the integrity of a leader. Frequently change the spirit of the main goals of the organization, constantly change the objectives of the teams and make a habit of course changes, not only harm the organization but also your leadership.

3. Cancellations and interruptions : If you want to prevent the team from feeling frustrated or confused about the roles that each person plays within their organization, it is important to avoid canceling the tasks entrusted to each other. Sufficient explanations. It is the same with the cancellation of projects without good reason or change the managers without notice. All this destroys the motivation of the whole group. It is the same to begin to do micro management within the organization.

4. Suspicion : The key quality sought by leaders is trust.
However, ruining this image is relatively easy: creating doubts or not recognize the achievements of the team, and that confidence will collapse.

5. Egocentrism : Leaders can be their own worst enemy. One should not let a selfish attitude destroy a good job that is done. Do not block the success of your group but take the
opportunity to congratulate and show interest in their work.
Likewise, you need to lend a hand when your team needs you.

6. Lack of membership: One of the great responsibilities of a good leader is to make employees feel that they belong not only to the team but also to the organization. The best organizations are those whose members are in the same direction, with the same motivation and purpose. If this sense of belonging is not promoted, the leader must know that he is destroying his leadership.

CHAPTER 25

THE SIX CHARACTERISTICS OF A
BAD LEADER

Not everyone is lucky enough to be led by
competent leaders. It is in fact quite possible that at
this time you may be shivering at the thought of
having to re-live your experiences with bad
superiors - or maybe your wounds have finally
healed and you are now in the company of a " A
person with an inspirational leadership style.
No matter if you were exposed to someone like
that, hired one or even you were one at one time or
another, the characteristics of a bad leader should
be identified in order to improve those dimensions
Of your business.
Here are some characteristics of a bad leader that

may seem surprising ... or a little bit too familiar.

1. Bad leaders avoid or ignore conflicts :

Whether in a direct conflict with another employee or when it is necessary to mediate between two parties in a dispute, a leader should not pretend that everything is going well and assume that things will resolve themselves, They themselves! Avoiding disputes or unpleasant situations can cause accumulated frustration, bitterness or miscommunication.

So, even if a leader might think he is serving everyone by avoiding confrontation, it can easily blow him up in the face one day or another. A good leader will be able to approach the situation with open-mindedness and a proactive mentality.

2. Bad leaders enjoy their power instead of empowering others. Being in a position of power does not mean that we have carte blanche to abuse this privilege. A true leader will make a conscious effort to inspire others, invest time in developing the team and help his employees become better.

When the title or position in the ranking rises to the head, the focus shifts from empowering the team to the desire to boost one's self-esteem. Leadership is not about controlling employees - it's about guiding them and giving them direction so that they can evolve and reach their full potential. The best approach is to clearly define the expectations and roles of each, to invite feedback and create stimulating opportunities for others to help them spread their wings. Empower employees to grow together!

3. Bad leaders never show their vulnerability

Leaders can perceive their role as powerful, robust and invincible and always want to demonstrate a mask of perfection - and this can be more intimidating than inspiring. What many leaders do not see is that this unrealistic image of power can be perceived as unaffordable or as having magical armor that protects these people from the everyday problems that employees suffer from day to day. Since when is vulnerability a bad thing? This allows those you are trying to hold accountable to see that you too are susceptible to errors, regrets and frustrations, and that criticism and rejection also have an impact on your state of mind - in short, that you too are a human!

Showing your vulnerability can make your subordinates understand you better and therefore feel less defeated by their own weaknesses. But as usual, moderation has much better taste!

4. Bad leaders are blind as to the strengths

of their team : I am not sure what category of defects one might classify this trait - perhaps as a bad judgment of character? The inability to identify the potential in others? In any case, you see what I mean.

Leaders who do not know how, or yet do not make the effort to see the forces of the workforce advance blindly. They can then delegate tasks to those not done to accomplish them, or simply fail to see the natural talent of an employee when it could have been very useful for performing certain tasks. A good leader uses tools that reveal the true potential of others, which not only allows for better assignment of tasks but also provides an opportunity to continue to motivate employees and communicate with them in the most effective way possible. Not to see these natural reflexes is to see only the tip of the iceberg.

5. Bad leaders never confess their faults:

Being a leader always means accepting a certain amount of responsibility for the outcome of projects or tasks that must be performed. This means taking the initiative rather than placing blame on others or feeling victimized. So we have to admit that the problem is our responsibility and take action to resolve or correct the situation.

A good leader does not just accept the praises for good things and ignore the bad ones. Of course, sometimes it's easier said than done! It is easier to point others and be passive, but it is much more productive to be proactive and want to improve the situation as well as improve oneself. The question then arises as to how one can improve one's performance and that of others. Like a captain who sinks with his ship, the leader should not abandon his team when the situation turns sour.

6. Bad leaders just do not listen : For leaders, there are several ways to listen. It's not about not only being silent when someone shares his point of view; it is also to pay attention to language non-verbal, to give comments to others about what they have just shared and to paraphrase and confirm what they have just said.

There is also the whole domain of self-knowledge, the true understanding of one's own style of communication. Maybe you are an extrovert who loves to have all the projectors branding on him and so start several discussions a day. Or maybe you know that during a discussion, you tend to go on tangents after 5 minutes. Regardless of the situation, know what you need to work on in terms of communication (how you express yourself and how you listen) so that you can hear what your team has to say.

It is very likely that a meeting with a bad leader was memorable for you. Maybe this person did not know how to handle a conflict (or did not even try), Which has escalated tensions. Maybe she was thirsty for power and failed to demonstrate her vulnerability, which intimidated her employees. It may not have been able to perceive the true strengths of its employees, has not listened actively or placed the blame on the others when she should have shown responsibility.

Regardless of which of these characteristics a bad leader you have had to find, enlist or participate in, the use of tools that can reveal such trends is essential for your organization.

Do you, or anyone you know, have what it takes to be a good leader?

CHAPTER 26

THINGS THAT YOU CAN NOT TELL
A BAD LEADER

Here is a list of the 37 secrets that customers
dealing with bad leaders have never told them face-
to-face . No matter what type of customers they
serve , no matter what business they are in, people
have these dispositions but will never tell to the bad
leaders. If you are a bad leader, imagine only for a
moment what your customers really think about you
... Imagine that you finally understand them after
you have gained their confidence, that you localize
them. As a leader, it is important to slip into the
head of your most loyal customers in the right way
and behave the way these secrets thoughts would

never come to their mind concerning you. By reading this list you will be more able to understand and would be able to better satisfy their most intimate needs .

1. No one needs to be perfect; Employees can not rely on it

2. employees can not tell him that they do not trust him anymore.

3. They can't tell him how happy and joyful they are every time he says "thanks".

4. They can't tell him that he isn't doing what he was committed to doing.

5. That he does not treat his collaborators as he would like to be treated.

6. Would he like to pay for his services if he was a customer?

7. Is it certain that these services are indispensable in society and judged to be an irreplaceable resource?

8. Do his employees understand all the messages he sends? That he should be clearer when he addresses them?

9. If someone comes to him and tells him that he has a complicated life, what could he tell him or do?

10. That people would like to feel special when they live his presence.

11. When he has money, would he like to give gifts to others ?

12. What would he like to do for someone who feels alone or taken care of ?

13. What does he think of those who think all the time that he mistakes ?

14. Does he like to receive small personalized gifts? Does he give gifts ?

15. Does he confess when he doesn't understand ? If not why?

16. What about unbalanced relationships?

17. What about jealousy when others receives more attention than him ?

18. Does he know how to deal with apologies?

19. Does he consider yourself the most interesting person? (Leader's Square)

20. What do you think when you sell or sell your services?

21. I want to buy your products but I need you to help me find excuses for myself.

22. Do you want us to think of you as a genius?

23. Who should do the dirty work? And who must reap the laurels?

24. Money is not a problem as soon as my obsessions are addressed.

25. You think you are good in one area? Well, you're wrong! Ask me and I'll tell you how you're the best.

26. They can't tell him that he should not pick up the phone while talking to people.

27. That he should not frighten or make people feel ashamed

28. That he is much lazy than he thinks ...

29. That he is too much selfish than he thinks

30. That he is very conceited that he wants to believe.

31. That he has less confidence in himself than he wants to tell others

32. That he is not as smart as he wants people to believe.

33. That he makes people think that he deserves more ...

34. That he should be brief and go straight to the point.

35. Most of the time, I do not know what I want, so I need you.

36. That he should not believe that everything wrong in his life is the fault of others.

37. That he sincerely believes that the world revolves around him alone.

CHAPTER 27

THE FIRST CHRISTIAN STEPS
FOR LEADERSHIP

Do you believe in your heart that God calls you to be a leader, whether in your community, in your family, or in the world? Do you want to be one of those who make this world a better place for everyone? Do you want to positively direct and influence others so that they can emotionally, morally and spiritually trust and trust in you? Make these biblical principles the foundation of your becoming as a divine leader.

Just a simple reminder: When God makes us leaders, we must have in mind that it takes time to become a great leader. Therefore, one must be patient in leadership. God always uses the humble, well-trained men, devoted to prayer, tried, disciplined and faithful to him. This means that God must approve you as one who will lead his people.

1. To think of one leadership position to serve : This kind of leadership is what God wants from us. God directs by servant leaders. Keep in mind that Jesus came to serve, not to be served and gave his life as a ransom for many (Matthew 20:28). To be a servant under Christ, one must first learn from him and offer his life for others on his model. And if you really do not know how to do it, pray and ask God to show you when and how to implement this principle. "... He who wants to be a leader among you must be your servant.

2. To forget the concept of boss: A leader who sees himself as a boss can not succeed in any social group. The only true leader, for both leader and the employees under him, is God. We all human beings work for Him. And if we do not want to be dismissed, demoted or unqualified for the work that God entrusts us, let us learn to be humble.

Humility means gentleness, patience, acting in gentleness and without anger or resentment. Accept that people sometimes look at you as weak; You may even sometimes feel embarrassed or morally hurt because of your humility. It is the way of God and his process to honor and promise you. God loves to work with the humble, because they totally depend on Him. I Peter 5-6 tells us, "Humble yourselves under the mighty hand of God, that he may raise you up in due time, and discharge him from all your cares, for he cares for you. *"If we want to be great leaders and receive God's favor on our lives, then we must be humble.* The position of "chief" must disappear;

It is pride; All that God hates. So practice humility.

3. No micro-management: When you delegate power to someone at the same time that there is a micro-management system in that same department, you aggravate your employees. Learn to delegate. This causes you stress, and allows you to focus more on what you need to do. Know that we all work together for God. So if you hire someone to do a job or fulfill a role, trust him; Do not do micro-management. Give him the opportunity and freedom to use his talents. You are not the master of all things. Only God is ... then let God work through those you lead to accomplish the mission. You will be amazed at the peace that will win your heart and the new relationships formed because of this trust.

4. Taking care of yourself as a leader : A good leader must take care of himself spiritually, morally, mentally and physically. I Timothy 4: 7 says: "... He should spend his time and energy in training himself spiritually. Physical fitness has some value, but spiritual exercises are much more important. We must study the Word of God, meditate on it, memorize it, and above all, do what it says. It makes no sense to go to church, hear messages, or read the Bible and not obeying what it says. It is deceitful according to Jas. 1:22: "Put into practice the word, and do not limit yourself to listening to it, by deceiving yourself by false reasoning.

NOTA BENE: The Bible does not say that physical exercise is not important. She simply says that the spiritual condition is much more important. We must therefore take care of our bodies as stated in Romans 12: 1 and 2 Timothy 2: 20-21.
We rest when necessary, practice self-discipline and self-control. In the end, the two exercises go together to help us continue the race of life.

5. As a leader, manage stress and emotions: Anyone who is a leader must manage both his stress and his emotions. As a leader in the image of Christ, we have the opportunity to better manage our stress and emotions. Jesus teaches us in John 15: 8: "In this my Father is glorified, that you bear much fruit; And then you will be my disciples. "Galatians 5: 22-23" But the fruit of the Spirit is love, joy, peace, patience, goodness, benignity, fidelity, gentleness, temperance; The law is not against these things. So you see, we have to drive with character.

Remember that it is during these moments of stress and emotion that the Lord looks at us to see how we manage them as well as his people as a leader. Let us be careful to strike the rock as Moses did in Numbers 20. We are "Joshua"! Our goal is to reach the promised land, not to accomplish it because of the people or uncontrolled emotions. Our duty is to lead the people of God by obeying them, and embracing grace, mercy and love, while keeping our emotions under control; It is under these conditions that God will be glorified!

CHAPTER 28

THE PARTICULARITIES OF THE CHRISTIAN LEADER

Jesus says that we are the light of the world, so we must be of those who take the initiatives, which show the way. Unfortunately, by laziness, tradition, false humility we refuse the place that the Lord would like to see us occupy, the place reserved for us. To emphasize what we have or what God has given us is not synonymous with pride but rather with gratitude to the Giver.

The world is looking for stars, leaders, models. Let us give them the opportunity to see us as the real models they need. The more we give what we have received from the Lord, the better we will receive what God promises us. For, says the Word of God, one will always give to one who has and will be in abundance, but to him who has not one will take away what he believes he has. Many refuse to engage as a leader, because they do not want to pay the price. The leader has different concerns than others. If there is a semblance of conflict between ministry and work, it is because God wants us to focus on work to support ministry, if at least we are honest with God.

What are the objectives of the Christian leader?

The objectives of the Christian leader are:

1- to inspire faith in the heart of the people it leads

2- to lead others to believe in themselves

3- to give love that fills a specific need.

You have to know that the world needs people who take initiatives. This can be seen everywhere. The big bosses or the executives of the companies call on coaches to boost the staff of the company.

1 Samuel 17: 41-54The Philistine approached little by little David, and the man who bore his shield walked before him.

The Philistine looked, and when he saw David, he despised him, seeing in him only a child, fair and handsome. Read more he world needs heroes, and the church also needs leaders who have succeeded (in all areas). The leader releases anointing, love and character. It also arouses or activates the faith of those whom it leads.

(The example of a CEO of a large company who visited one of his sick workers. This visit quickly raised the morale of the patient and created a positive effect in the output of the workers).

2. The leader always communicates a vision. It is a catalyst that leads to a goal, a vision. Even in the desert, in the bush, he sees a path to follow. The leader comes to discover the talents and

qualities that others have around him.

If it is the intercessor, he must refrain from manifesting any authority and influence on the leader. Its part being to cover with prayer its spiritual conductor, and to raise around it a spiritual wall. It should not try to control the leader in one way or another. He must never try to convey his vision or impose his vision on the leader. For if a well-positioned soul can be a weapon for the kingdom of God, a badly positioned
soul is against the kingdom of God.

3. Victory in conflicts of influence : Conflicts are manifested everywhere where beings organize, work towards the realization of certain projects. Conflicts reveal the vitality of the group, they are not aroused by inexperienced people, but rather by the elders who think they know a little more than the established leader.

Therefore, the leader must make sure to keep his place of leadership by highlighting the vision. He must know how to keep the vision and authority received. Any action or word tending to minimize, to trivialize, or worse to despise authority, is of consequence. However, if for several reasons the leader is mistaken, the collaborators must resort to prayer and never defy or confront it.

1 Timothy 2: 1-3 I exhort therefore, above all things, to make prayers, petitions, petitions, and thanksgiving to all men, to kings and to all who are raised in dignity, So that we may lead a peaceful and quiet life, in all piety and honesty. Galatians 2: 11. But when Cephas came to Antioch, I resisted him, because he was reprehensible. If, at the end of the prayers of intercession in his favor, no transformation takes place, the collaborators are free to leave the spiritual

conductor. What if the leader is confronted with cases of revolt or rebellion, as was the case of Coré, Datan and Abiram in Numbers 16: 1-35? The offended leader must seek the direction of God, as Moses did. He must not rush to make decisions without receiving instructions from the Lord. He must then approach the rebels or the rebels to bring them back to the goodwill. For it is possible that the leader negligently entrusted the entirety of his authority to his deputy. In this case, it often happens that the two fight to keep the power or the control of the group. The leader will always need to consult God to know his direction.

But, in any case, it is more important to avoid struggles for power than to win them. The leader must teach others under his authority the rules of delegation of power. He must entrust power only to those in whom he trusts. The person in charge of the interim must know the limits of the authority entrusted to him. Thus, he must constantly recall the mission entrusted to him, say the name of the leader of the organization of which he assures the interim in his absence. However, the selected person must not take possession of the group. He must make a regular report on the group's evolution to the absent main leader. He will give him instructions.

Anyone who has been appointed to a group by the Lead Leader must respect the group's constitution. He must study the group patiently. He must avoid making or causing comparisons between him and the one for whom he is acting.

He must make sure that the members of the group do not make comparisons between him and the absent main leader.

The leader must, in any case, avoid quickly naming people. It must organize itself to be productive. He has to make everyone work. The whole group must be subject to its authority. Each employee of the leader must position his soul in relation to his domain.

The leader must be persevering. He must not abdicate his task unless instructed by the Lord. Clearly, the leader must focus on the following ideas: The leader must be sure of himself. He must be certain of the vision and his anointing and know how to communicate it to all.

5. The leader must be legitimate in the exercise of his authority when exercised within his sphere of influence.

2 Corinthians 10:13," For us, we do not want to boast out of all measure; We shall, on the contrary, take as a measure the limits of the division which God has assigned to us, so as to make us come also to you.

Romans 12: 3By the grace that was given to me, I tell each one of you not to be of too high opinion of himself, but to put on modest sentiments, according to the measure of faith which God hath departed To each. It must not imagine its sphere of influence, but it must be realistic in its concrete appreciation.

6. He must be conscious of taking steps in the exercise of his authority. And it must not through our heart we can bind or loose souls but we must do so within our sphere of influence. (2 Cor. 10). John 17:12 When I was with them in the world, I kept them in your name. I have kept those whom thou hast given me, and none of them was lost except the son of perdition, that the scripture might be fulfilled.

10. The leader exercises a spiritual blanket on the sheep by taking these sheep to heart. But a sheep also exercises a
spiritual covering on itself. This spiritual cover is also a reality in the family unit (for the father and the mother). To

be continued.

11. One recognizes a true leader more to what he does not do than what he does. Listen to their fears Leaders are not supermen, like everyone else they have weaknesses and limitations. But, unlike others, they know how to ignore and realize their dreams despite everything.

b) Not Challenging To follow its own path and not rely on others does not mean not listening to the wise advice of competent people. Great leaders know they are not blinded by their own vision.

c) Trying to be what they are not The difference between someone who knows success and a real leader is that he never tries to give himself a genre or to appear what he is not Great leaders know that authenticity is much more effective
than self-promotion!

d) Do what everyone does. True leaders tend to naturally not follow social codes or group thoughts and challenge the status quo.

e) Fear of competition Great leaders derive their motivation from overcoming competitors. They are confident in their ability to win the battle of competition.

f) Losing time The great leaders do not lose a minute because they want to move quickly in the realization of the vision they have for their company. They do not procrastinate and do not waste time wondering what others think.

CHAPTER 29

THE QUALITIES OF A SPATIAL LEADER

Our society is becoming more and more complex. The traditional methods of managing family affairs to social and non-profit organizations need people who have some form of special training and especially who see Jesus Christ as his model of leadership.

As you read this chapter, we are not someone specially trained in leadership, but someone who sees emptiness in the whole circle of life especially in Africa. From government to family affairs going through churches and businesses, we see how leadership is almost something unknown. The consequence of this state of affairs leads us to think of our countries marching towards the direction of another Black country, the third nation in the world to become independent in 1801, namely Haiti. I do not see anything African countries that duplicates the Haitian experience in 200 years. All the characters, attitudes, principles, values and personalities we cover in this book are unknown, not existing among the organizations and systems in place in all African nations. We are certain that this book will be rejected, as it will be the African fuel of political, economic and social values and lifestyles. God has placed us all in a leadership position, if not in our workplaces or churches, then certainly in our homes as parents. I know there have been times when I do not explain the qualities of a divine leader. I hope to learn by writing this book.

This chapter is filled with leadership lessons. Below are nine principles that are essential characteristics of a good, divine leader.

1. A good leader seeks the direction of God:
Is there anything more important in a leader than he
seeks the guidance of God? Proverbs 16: 1 says,
"The plans of the heart belong to man, but the
answer of the tongue comes from the Lord."
Verse 3 adds, "Commit your work to the Lord, and
your plans will be established. "The heart of man
plans his way, but the Lord establishes his steps." A
good leader seeks the Lord, surrenders to the Lord,
and the Lord sets the next steps.

2. A good leader is modest, not arrogant:
We all met the leader of know-how, the leader of
the "submit-or-other". But Proverbs 16: 5 says,
"Whoever is arrogant in heart is an abomination to
the Lord; Rest assured, he will not go unpunished. "I
do not know about you, but I certainly do not want
to be mentioned as an abomination to the Lord. It's
a pretty scary thing.

3. A good leader is a peacemaker: Proverbs 16: 7 says, "When a man's way pleases the Lord, he even makes his enemies at peace with him." Yet many leaders are not interested in examining an opposing viewpoint,

Other ideas. We have lost the ability to sympathize with others, and compromise has become a bad word. There is something to be said about respecting the principles. I believe God calls us to be firm. He does not, however, call us to be jerky. And when our "audacity" is interpreted as "coldness," we do not do it properly.

4. A good leader is right and right: "A little justice is better than great revenues with injustice" (Proverbs 16:

8). I believe in goals, and work hard to achieve them.

But, the end always justifying the means is simply not true. A good leader is more interested in doing things the right way.

5. He surrounded himself with honest and trustworthy advisers .. then he listened to them. "Straight lips are the joy of a king, and he loves him that speaketh righteousness" (Proverbs 16:13). Do you know leaders who surround themselves with people

"Yes"? Personal insecurity prompts them to seek only positive reinforcement for every decision they make
Take. An intelligent leader is surrounded by smarter people, who are willing to speak their mind and offer sound advice. After all, "Without counsel, the plans fail, but with many counselors they succeed" (Proverbs 15:22).

6. A good leader is a good student: Proverbs 16:16 says: "How much better to have wisdom than gold!

Understanding, one must choose rather than money. A good leader should always be learning, growth and improvement. The day you feel that there is nothing to learn, it is the day when pride and arrogance have taken root. And, we have already discussed how the Lord feels about arrogance.

7. A good leader is humble: We have seen countless prominent examples of Proverbs 16:18: "Pride goes before destruction, and an arrogant spirit before a fall." From politicians and celebrities to CEOs and pastors, many Have seized the titles as their empires have fallen. In most of these cases, it is pride that slipped in. They considered themselves invincible, but quickly discovered that no one is. "It is better to be of humble spirit with the poor than to share the spoil with the proud" (Proverbs 16:19).

8. A good leader is reasonable and kind:

"Common sense is a source of life for the one who has it, but the instruction of the insane is madness. The heart of the wise makes his word wise and adds persuasion to his lips "(Proverbs 16: 22-23). Being intelligent and responsive makes a good leader more convincing and effective. A good leader uses "words of grace" (verse 24), not a word that is "like a burning fire" (verse 27).

9. A good leader is slow to anger: We have all seen the Cartoons in the movies and television of the angry boss; The person who cries out without reason, barks orders and reprimands and demoralizes the staff. Perhaps you have even worked for such a person. The Bible says that "the one who is slow to anger is better than the mighty, and the one who directs his mind is the one who takes a city.

By reading these qualities of a good leader, I hope you find them as difficult as I am. God tells us how to be effective, the rulers of God. It is up to us to put aside our human tendencies and embrace these principles. It is also up to us to pray for those under whom we serve, that they would also be the good leaders that God wants them to be.

CHAPTER 30

THE TWENTY-ONE (21) IRREFUTABLE LAWS OF THE LEADERSHIP. On John C. Maxwell

For John C. Maxwell, whether it's success or failure, it all depends on leadership. As you work to build your organization, remember that ... Staff determine the potential of the organization. Relationships determine the morale of the members of the organization. The structure determines the size of the organization. Vision determines the direction of the organization. Leadership determines the success of the organization.

To develop your leadership, John C. Maxwell wrote these 21 laws. This summary will, I hope, encourage you to read the entire book ...

1. The law of the lid: Everything you seek to accomplish will be limited by your ability to be a leader of men. The talent of the team is rarely a problem ... the leader and the key players make the difference. The higher the leadership, the greater the effectiveness. The greater the impact you want to create, the greater your influence must be. Recognize your talents and limitations.

2. The law of influence: To be a leader is to exert influence, nothing more, nothing less. The influence is linked to a single clear, clear and mobilizing objective. Influence is about action and the ability to make things happen. Influence helps to put things on the agenda for the right reasons. The influence is not related to your title, it is gaining in action and you have to work hard to get there.

3. The law of the process: The leader develops his talents day after day. It distinguishes leaders from ordinary mortals. The leader has a long-term vision to grow his assets. The leader develops his awareness and efficiency. Being a leader tomorrow means learning today.

4. The Law of Navigation: Anyone can bar a ship but to chart the course it takes a leader. The navigator has a vision of his destination, he understands what this will require and he knows who he needs to get there. The browser knows that other people depend on it and its ability to chart a path. The browser listens to what others have to say. It is not the dimension of the project that determines its acceptance, its support and its success, it is the dimension of the leader.

5. The law of E. F. Hutton: When the real leader speaks, the people listen, it is he who holds the power not only the post. The leader develops in seven key areas: he is, he knows, what he knows, what he feels, what he has experienced, what he has done and what he can do. True leaders know how to use the leadership of others.

6. Firm ground law: Trust is the foundation of leadership. People know when you make mistakes, needless to hide them. To develop trust, you have to develop your skills, work your relationships and have a good character. Character makes trust possible and trust makes leadership possible.

7. The law of respect: People naturally follow leaders stronger than themselves. If you are respected as a person, you are admired; If one respects you as a friend, one loves you; If you are respected as a leader, you will be followed. True leaders see and respect the leadership of others. Respect is gained by telling the truth and by investing quality time with others.

8. The law of intuition : Leaders evaluate everything with a bias towards leadership. A leader evaluates the situation and instinctively knows what to do. Intuition develops with a multitude of experiences. Intuition allows us to see what is possible. Leaders who want to succeed maximize every asset and resource they have for their organization. Without intuition you risk having blinkers, it is one of the worst things that can happen to a leader.

9. The law of magnetism :The ones you attract are consistent with what you are. What you get is determined by who you are. If you think your collaborators can be better, it's time for you to be better. People are attracted to leaders who have values similar to theirs.

10. The law of contact : Leaders touch hearts before asking for help. The strength of the contact is to make sure that the other feels happy to be with you. To lead yourself, use your head; To lead others, use your heart. The stronger the relationship and the contact, the more people will follow you. A great leader standing in front of a group of 40 people sees 40 people with aspirations, each wanting to live and do good. The leader has an obligation to take the first step to establish contact and maintain it.

11. The law of the close circle : The potential of a leader is determined by those who are closest to it. If someone leaves, look for someone better to replace it. The leader reaches strength within a group and helps people find strength in them. Invest your time with your best assets and delegate as much as possible. Keep in your circle those who are able to raise others and help you improve.

12. The law of delegation : Only the self-confident leaders delegate their powers to others. To delegate you have to believe in others. People's ability to achieve things is determined by the ability of the leader to delegate. If you do not delegate, you create obstacles inside the organization. The delegation of power allows the arrival of constant changes, because it allows to develop and to innovate. You acquire authority by giving it to others.

13. The Law of Reproduction : Only one leader can generate another leader.85% of the leaders have become thanks to the influence of other leaders. To grow, spend time with other leaders better than you. Create an environment where leadership will be appreciated by launching a vision, providing motivation, encouraging creativity, allowing risks and empowering.

14. The Act of Accession : People adhere to a leader, then to vision. If you feel that the messenger is credible, then you feel that the message has value. When people do not like the leader or the vision, they look for another leader. For people to join, they need to understand you, understand your hopes and dreams. A vision is not necessary.

15. The Law of Victory : Leaders find ways to win their teams. Great leaders accept nothing less than victory. It is the training that allows victory, not just talent. When the pressure rises, the great leaders are at their best, all their forces rise to the surface. There are three components to victory: a unified vision, diversity of talents and a leader dedicated to winning and optimizing the potential of its players.

16. The Big Mo law (or the law of impulse)
The impulse is the best friend of the leader. To

create the impulse, you need preparation and motivation. When you have the impulse on your side, the troubles are short lived. When there is impulse, people are motivated, inspired and want to raise their performance levels. To keep the momentum, we must focus on what is possible rather than on the impossible, we must celebrate successes however small.

17. The law of priorities : Leaders understand that acting is not accomplishing. Doing more does not mean that we are doing better or that we are fulfilling our mission. Effective leaders organize their lives according to three questions: what is required, what gives the best result, what is the best reward and stimulates you the most? The priorities must be constantly re-examined and reorganized. Success comes from concentrating its collaborators on things that matter.

18. The Law of Sacrifice : A leader must be able to give up to show and seize opportunities. You can not ask for sacrifices if you do not. As you get into leadership, responsibilities increase and fees decrease. The higher the level of leadership you want to achieve, the greater the sacrifice.

19. The law of the moment : It is also important to choose the right time to know what to do and where to go. Bad action at the wrong time leads to disaster. Good action at the wrong time provokes resistance. The wrong action at the right time is an error. Good action at the right time leads to success.

20. The Law of Explosive Growth: To add to growth, lead

disciples, to multiply growth, lead leaders. Leaders who form leaders want to be succeeded - they focus on their strengths, treat leaders as full-fledged people according to their impact, delegate power, invest time in others, grow by multiplication And have an impact on people even without knowing them.

21. The law of inheritance: The lasting mark left by a leader is measured by the succession he has left behind. The true leader creates a culture of leadership. The true leader favors team leadership to individual leadership.

MAXWELL, John C., The 21 irrefutable laws of leadership: follow them and others will follow you, Saint-Hubert, Quebec, International Publishing and Dissemination Group, 2002, 227 p.

CHAPTER 31

LEADERSHIP RULES ...
According to Dan McCarthy

Dan McCarthy, director of leadership and business development for a Fortune 500 company, recently shared in his blog his 10 golden rules of leadership.

1. Take your responsibilities to heart : Never take your leadership role lightly, as everyone relies on you. Your mistakes can be detrimental to the lives of employees and members of the community. Take responsibility for your actions.

2. Develop the potential of subordinates : The leader is responsible for developing the full potential of his employees. Its duty is to offer stimulating work to the employee, which is what allows the employee to grow.

3.Create an AAA team : You are responsible for creating a AAA caliber team. Your goal is to hire, keep, and promote the best employees. Your goal is also to bring all members of your team at AAA level. Your team must work hard and its level of performance must remain superior to others.

4. Be strategic : It's your job to ensure that your employees do a strategic job. The goals and activities of the team must be directly related to the objectives of the company. Each team member must also target individual goals and objectives as part of a training and development plan.

5. Valorize the meetings : It is important to value team meetings and individual meetings. They represent the best way for the leader to demonstrate good leadership.

6. Be a role model for others : Your own success must be as important as the success of the team. Be a role model for your employees.

7. Use a positive approach :
Be positive when you talk about the company, employees, products and services, customers, your goals and departments. Constructive criticism is always preferable.
Never give cynical or sarcastic comments.

8. Support the team : Be supportive of your team, despite the debates and challenges to be overcome. You may or may not agree with some decision-making, but you are part of the team and have to support it.

9. Treat your employees as adults : Your employees are responsible and intelligent adults and they should be treated as well. Respect them and do not abuse your leadership title. Do not mingle with all decisions and do not tell your employees how to do their job.

10. Listen! Fight your natural instinct to evaluate and reactImpulsively to a situation. Listen and ask more of questions. You can then better judge a situation and react accordingly.

CHAPTER 32

THE THIRTEEN (13) LEADERSHIP RULES

It is Colin Powell, the former US Secretary of State and former boss of GI's during the Gulf War. Here they are, adapted to the times of social media that have become a true vector of leadership.

These are rules of leadership and, as you will see, all have something to do with the story we tell, our storytelling. Apply them, even if they sometimes seem a bit too American. And then your storytelling will truly be labeled leadership.

Rule # 1: It's just as catastrophic as you think. It will have a different pace tomorrow. Reacting quickly to avoid virility is also likely to react badly and therefore make things worse. The first rule, therefore, is to give time to a task of assessing the situation before acting-react, because this situation, ultimately, may not be as catastrophic as it seems. And the story you tell will be more coherent.

Rule # 2: Go mad and then overcome it. Of course, we want to put all our emotion into our reactions, especially on social media. A certain balance must be maintained. Especially since in general, a negative publication has a very limited lifespan. Excessive anger and emotion can only prolong debates without much interest for much longer than necessary. So be crazy (keeping this private madness) and then act to turn this negative situation into a neutral or positive situation.

Rule # 3: Avoid leaving your ego so close to your current position that when you lose your position you also lose your ego. This is about reinventing oneself. And for that: being integrate these ideas into what you do, have a real desire to constantly have a competitive edge.

Rule # 4: We can do it.
Well, it's just that we have to show the positive side of change, when we are a leader in charge of steering a change. That is to say: answer positively to the question of its teams on the "why" of the change and "what is the effect on me?"

Rule # 5: Do not be deceived by anyone.
Friends, allies or enemies: choose the right people to take on each of these roles. The key questions are: who, why and how you interact with them. Do not engage with people or organizations without asking yourself the question: who influences, how

and why?

The why is probably the most important question.

Rule # 6: Do not let facts get in the way of a good decision.

Rule # 7: You can not make decisions in place of others. Do not let others take it for you. Be sure to consider three things: understanding the key issue of the challenge you are managing, studying the different options from all angles and perspectives, asking a lot of questions (you will understand the options better).

Rule 8: Check the details.

Rule 9: Share honors. With those who deserve it

...

Rule # 10: Stay calm. This will save you from saying things that you might regret afterwards. It's a bit contradictory with a rule mentioned above

Hey, who said leadership was something simple!

Rule 11: Have a vision.

Rule # 12: Do not ask for advice from your fears or detractors.

Rule # 13: Permanent optimism multiplies your strengths. For enthusiasm is contagious.

CHAPTER 33

THE FIVE (5) GOOD ATTITUDES OF A GREAT LEADER

1. Be generous : The law of communication is the law of exchange. In this exchange, you are interested in others, if you want others to be interested in you. Be generous, give and you will receive in return.

2.Respect others and you will be respected in return. This is the principle of reciprocity. Your benevolent behaviors will impact you. Find all opportunities to increase another person's self-esteem. Your associate will feel valued and your self-confidence will be enhanced.

Do not forget that self-esteem, feeling respected, is

a fundamental need of man, described in the pyramid of Maslow's needs. Take advantage of every opportunity to express kindness, show others that they are important, that they have value. You will be rewarded in return.

Adopt a positive attitude in your relationships with others isalso be quick to listen.

3. Listen to your interlocutor : In a conversation, be concentrated. Listen to your interlocutor carefully, avoiding thinking about the answer you will give. It's hard for everyone. We all want to give our opinion on the spot. But listening attentively, it is tempting to understand well, to hear what his partner says.

By feeling listened to your interlocutor will feel valued. Leave it room, give it importance, let it express itself. Do not judge him. Have an attitude of benevolence so that he feels comfortable expressing his own thoughts.

This applies in both areas that govern your life, work life and personal life. In sales, for example, if you listen attentively to your customer's needs and respond effectively, your customer will be happy.

If you plan a project with one or more people, listening to the other allows you to have all the necessary information on their needs so that there is agreement between the different parts of the team and succeed to reach goals.

In a couple's life, understanding the other is above all listening to him, listening to what he has to say, listening to his needs. To listen to him is to feel respect and a way to make him understand that it is important for you. Your spouse will return it to you in turn having the same attitude of openness towards you.

4. Be open : Some people are not very open to

others and do not wish to build relationships with people who do not resemble them. It is a shame because it cuts off the riches of human diversity. Make an effort to meet others. At work, get to know your colleagues, converse with them, join jokes. In short, multiply the exchanges.

As soon as the opportunity presents itself, help others as much as possible, by simply being attentive to their problems, being useful for a record, being useful in general. You will reap the benefits in the short or medium term.

Use the smile abundantly as it is an excellent icebreaker. At work, this shows that you are accessible, non-hostile and will want to meet you. In addition, it relaxes the atmosphere.

5. In the event of conflicts : Unfortunately in all relationships conflicts are inevitable.

At the office it is enough to have a divergence of point of view on a file with our colleagues. With our friends, our spouses, there may be some tensions.

Again, attentive listening is important. Try to put yourself in the other's place. This will allow you to understand what has

crumpled the other and will allow you to explain yourself.

Do not be defensive. Look for what is true in what the other says, which can be valuable information to give material to your answers. I still have to be on the defensive and want to win at all costs. But it is more rare than before and I assure you that since I have been listening and no longer centered on my little person, my relations with others have become healthier, more peaceful.

In order to defuse conflicts, it is best to try to find solutions that meet the needs of everyone.

Sometimes it is difficult to continue a discussion if all the protagonists get carried away. It may be necessary at this time to take a break to calm the spirits.

If your partners are in a communicative construction logic, do not hesitate to come back to the discussion. Cultivate the art of listening, hearing, saying and sometimes not saying. You can not be a good communicator all the time, but if you apply many of these principles, your human relationships will be greatly improved.

CHAPTER 34

TEN (10) GOLDEN RULES FOR
LEADER AND TEAM

Do not promise more than what you are sure to offer. Stay humble in your promises (but not in your goals!), Even if you are almost certain of the success of your professional projects or the achievement of your goals, say the minimum. However, in your heart, be sure you will achieve these goals and ambitions, but do not share it with anyone before anything has been signed, validated or officially approved. It is better to give the impression that you are always doing more than you promise that the opposite. For in the image that you convey, someone who does exactly what he says is a sure value, just as someone who fails to fulfill his commitments seems to be unreliable.

2. Do not expect to receive all promised by your hierarchy. Do not be waiting for all the promises made by your hierarchy. Often to bait you, help you reach your goals or get you where they want, your hierarchy Promises you better working conditions, a promotion, an increase in salary. They tell you that it is virtually validated. You believe in it. Then they slip that to get there you have to show that you are good,

Responsible, consistent in performance. Yet you are. It is a way of subjecting you to greater investment. They give you a goal, and at the same time they get you there because they are the only holders of the decision-making power of their promise and the appreciation that the goal is reached. To avoid continual expectation, and to be completely subject to demand, imply that you know your value and that you have other proposals elsewhere. While saying that you are passionate. Because if you give to see that this technique, because it is a technique, walks with you, you will have given them the keys to make you advance in a way very constraining for you.

3. Be independent in your function as much as possible. Of course we depend on other services, other sectors, our pairs, our team, but your task to you, why you have been hired and trained, be clearly assured In order to be as much as possible in known terrain. You have to expand your knowledge and be versatile to the maximum, not to need others in the process of your job. Let me explain. You are web marketing. You create sites for companies for financial purposes. A Croatian textile company contacts you to develop its international aspect. You have to find out what makes Croatian culture, textile, and their assets so that you can best represent them on the web without the need for someone to do this work for you. Because the more you know about your customers (their culture, their view of the world, etc.), the closer you get to their expectations and goals. While if you ask your assistant to write for you a report on their activities, textiles in Croatia or other, you will have a filtered relationship through her eyes. Nothing very specific.

4. Become aware of the importance of communication. Everything is always about communication. Conflict, misunderstanding, bad negotiation, bad relationship ... everything finds its solution in a communication understood and used wisely. It is not about being false, but about being clear with what to say and how. You must get into the habit of communicating and not talking. The art of inducing, the art of convincing, the art of collaboration are all derived from good communication.

5. Be transparent to all the facts : Never give the impression of being a custodian, or keep information common to you, because you would induce suspicious behavior in the other. And above all, in a position of leader know how to account for the ins and outs of the goals related to the company. Does your company have a humanitarian foundation? Keep track of your successes both in the commercial sector and in the secretariat. The more you convey a unit in your sectors the more you will unite your teams in an efficient link. This rule is also self-evident for the teams.

You do not have to report on your activities to you, but be factually accountable to your hierarchies when they ask you to do so. Figures, dates, clients, yes. But the nature of appointments or your opinion about these clients as well as your managerial and marketing techniques are your personal baggage to you and do not fit into the box of what needs to be said.

6. Leader, be an example of what you are asking for:

You are a leader and ask for transparency? Know to prove that you follow your precept by being also transparent, but in a way adapted to your position. You are general manager

---

9.Have strategies to tame your anxiety-causing thoughts. We all have moments when the anguish overwhelms us: before a meeting, after a conflict, in the management of a crisis ... and it is in those moments that one imagines oneself

in spite of oneself the worst issues . The loss of confidence in one's own footsteps follows. To avoid this, as soon as the process of anxiety begins, stop (even if you have to go to the toilet for that) and imagine the positive result of this stressful situation.

10. Visualize the next step: the path you will take to your office before meeting your collaborators, your energetic handshake when you reach out to the one with whom the conflict is emerging. In short, visualize only the positive outcome of your dilemma. Do this as many times a day as your anxieties return.

Thinking about your positioning is a tool that promotes professional success because, like any awareness, you become master of your choices, and actor of what you want to create.

CHAPTER 35

13 PROFESSIONAL RULES OF DIGITAL EVOLUTION.

Julien Lucas Solo taker, blogger, designer, bizcampus.co.uk

OFFICE LIFE: The world has become uncertain, the world of work is changing, its rules too, and many people no longer know how to manage their working lives. We want to explain a little bit about these new rules.

We live a digital revolution, power is now accessible to individuals. We are moving towards a more exciting, more exciting, more humane world of work. Finally, for those who want to take the trouble to get there, they must learn to live by other rules.

1. More than ever, we are responsible for our own lives.

No one will come to save you. Your business will not care for your career. Your success in business will depend on your ability to create connections. Do not count on anyone else.

2. Work is organized as a result of the projects.

Forget promotion, it is a practice of the last century. Today,

some barely out of school, have already changed ten times of job in 3-4 years. You will have several jobs during your career. You should think about what combination of skills, creations, achievements you want in your "portfolio" in ten years through the different jobs and projects you are going to experience.

There is no shortage of possibilities, you can be employed, contractual, freelance, business owner, startup founder, social entrepreneur, independent producer, working in a non-profit organization or even simply being an internet personality with your channel YouTube.

3. The world of work is being tailored for risk takers. Job security is dead, there will be circumstances that will require you to change jobs several times. The new security is in your network, in your skills, in the diversification of your sources of income, it is in the new web tools that you use. And failure no longer means that you are dead. Today you can start a business on the internet with minimal funding. With digital, the impossible becomes possible. Surf on fear.

4. We are in the era of "next door" projects

A project next door if you are an employee is a great way to test and try new business ideas. This may also be a backup

plan in case you lose your job. These next projects do not only apply to employees, they can also be a good backup plan for business owners when the market moves or sales are low. I thought I would succeed with the product sale on my blog, I did not succeed, so I launched my freelance designer career, and it succeeded me more. This short-term project allows me to finance my long-term project.

5. Continuing education (training becomes essential :

Industries change quickly, and the moment before becoming obsolete can also happen quickly if you do not train. Education does not definitely stop after school. The school is not able to train you for the rest of your life. We become responsible for our own education.

6. Take the job for skills to earn more than for money :

We have families to feed; We need money to support ourselves. However, wages increase very little. If you have the ability to defer instant gratification, to endure frustration, take a job that will enable you to develop your skills. This will make a difference in your life in a way that salary and your title can not because the skills development can fundamentally change how you operate and what you have to offer.

7. Employers begin to look less at diplomas:

Employers look at what you have done, with whom you have worked, they look at your ability to solve problems, find new ways of doing things, your ability to live in a world dynamic.

If you are without a diploma, and can share your specific knowledge, then you are demonstrating your value to others. Creativity becomes the new literacy.

8. Many ordinary jobs are doomed to disappear in the next decade. With the computer, some jobs will be totally replaced, not just in the industry, even office workers are at risk and could be replaced by robots, artificial intelligence or automation.
Make the ant more than the cicada, plan for the future, develop your talents and skills with high added value today.

9. The new dream is to have time : The dream is not so much to be able to afford a house, it becomes less and less possible given the professional uncertainty, the new dream is to have time. We will not do better than our parents, at least financially.

The dream of the new generation is mobility, the flexibility to do one's own job and the ability to build a career that is the expression of who you are as a person. The new generation dreams of entrepreneurship.

10. Define what success is for you : Everyone will become responsible for his own success. There will be no tracks to follow or career ladder to ensure a career path. The worker of tomorrow will have to be smarter than his predecessors.

This is already the case, today's generation people take risks that the older generation would not even have imagined. Define success according to your roots, your values, your strengths, how you manage fear and doubt, how you work, your creative life, the quality of your life, your relationships and collaborations, and your physical and emotional well-being.

11. The future of the world of work belongs to the self-employed. Some are for developers, others for those who want to work with large companies or agencies. There are freelance agencies whose job is to find you work.

12. Your business success depends on your connections.
No one succeeds alone, to succeed in business, everything is in the art of creating connections.

13. To have the job tell a good story : And your story is widely told by Google. Like it or not, Google tells a story about you. Employers will choose someone who tells a good story, and they will start by going on Google (a good reason to have your web presence).
It is difficult to control one's own history. Yet the jobs are won and the business is done based on the strength of the story told on Google.

What have you done? Who are your connections?

Who is connected to you and your message?

What do you think? What is the thread between

your jobs?

You have to pack, illustrate and shape it into stories.

CHAPTER 36

THE FIVE (5) LEADERSHIP PRINCIPLES TO SUCCEED.

We all have a vital need for successful projects. But what exactly does a successful project mean? Is the success of the project successful delivery, on time and in budget? Or is it the way to succeed? Do the results always matter the most? What else but the successful project? And what does it take to achieve success on the project? Does success fall from heaven? Is it limited to a few lucky ones who are in the right place just at the right time? Is it a coincidence? Or can we really plan success?

There is no doubt that good project management is a critical success factor. That is, a project can not be executed without project management, be it formal Or informal. You must have something that ensures cohesion. Underlying is the assumption that we have need some sort of order to organize and execute a project. Someone has to do this something. In this sense, project management helps to lay a framework, providing structure and direction to potential chaos. Without this structure, a project does not lead anywhere; He will most likely fail, if he ever manages to take off.

If you want to produce results of this apparent chaos, you need to set up the structure that allows creativity, innovation and results. Project management provides excellent tools to build this structure. They are important and necessary for project success. But are they sufficient? I do not think so. In fact, I claim that unless you adjust them in the right direction, they remain ineffective. If you really want to secure the success of your project, you need to understand what to do to give the right direction. Project management alone will not succeed. It requires leadership - your leadership. Without project leadership, there is no direction in project management. Leadership is the decisive factor in improving the chances of project success. Therefore, effective project management must be based on solid foundations in project leadership. Without leadership, there is a good chance that this project will

"Than another project."
Based on my own experience in project management and literature review on leadership, project management, business, systems and complexity theory, I have identified five simple yet powerful principles of leadership that, If applied consistently, can you lead your project to success. The five principles of leadership for the success of the project are:

1. Building vision

2. Nurturing collaboration

3. Promoting Performance

4. Cultivating Learning

5. Ensuring results

Let us take each principle one by one.

Principle 1: Building Vision: Building vision means having the same understanding Monitoring of progress towards this vision is a key factor in success of a project and a team.
A project vision gives the complete picture of your project. The objectives of the project qualify this vision, make it specific. Both the vision and the project objectives are crucial to project success. Together they define the direction and set the tone for your "project trip". They **Complement each other.** The vision inspires your journey. It defines the purpose of your project.

The key to building the vision is that people need to be able to touch the finger of vision in their daily activities. Give them the chance to identify with the vision. Involve them in building this vision and participate in making it a reality. This helps to build the relational and necessary membership of these people to carry out the project. Make it fans of vision. Let it constitute their motivation and passion. Let them go wild.

The story of a visitor who was curious about a construction site illustrates the power of a shared vision of the project.

This visitor approached a group of workers to learn more about construction. The first worker replied that he was laying bricks. The second worker

Said he was building a wall. So he asked the question

Principle 4: Cultivating Learning : As humans we all make mistakes. Effective leaders encourage their teams to explore new avenues and make mistakes and learn from them. An effective leader incorporates enough time for the team to learn, create and innovate.

As the project leader, you serve as a partner and coach to study and share information. You facilitate learning. You are not the only source of information. Instead, you create a learning environment for your team. Explain the expectation you have of each one in your team that he joins you and supports you in the culture of learning for the purpose of the project.

Learning is not a one-time activity, say, in the form of pre-the beginning of your project. This is ongoing and should be on your team's daily agenda. Establish regular sessions with your team where you review

Past performance, share information about planned accomplishments, address and resolve barriers together. Invite external reviews. Outside views offer different perspectives; **Fresh perspectives** and
Not biased. If they aspire to help the team identify risks and unknown issues and overcome them, these external project reviews can be a great opportunity to learn.
When you or your team make mistakes, learn from them. Correct your defects, improve your performance and **Continue to move towards the fulfillment of the vision of the project.**
Cultivate learning from the beginning
project. This significantly increases the rate at which your team can run and support performance all the time and thus secure delivery.

Create a space for your team members to be creative, try something new, share their ideas and learn from each other. Allow enough time for your team to think outside the boundaries, beyond known paths and find new avenues to achieve project goals. Allow your team to perform, make mistakes, learn and innovate. This reduces the uncertainty

Information flows more freely. Team members are not afraid of making mistakes. They see mistakes as opportunities for learning and they

Help each other to solve problems. As a corollary, if you want the performance to bring the results desirable you must cultivate learning. He can not to have sustainable performance without learning and there can be no outcome without performance.

Principle 5: Ensuring results :

Delivering results is a need AND a result of effective project leadership. The delivery of the project is a team effort, not an individual effort. The effective project leader builds and guides the team to deliver results in incorporating the first four principles of leadership.

Ensuring results is not just about end results. No more than project success and project leadership. The fifth principle tells us that in all our activities we keep in mind the project vision and produce the results that benefit the objective of the project. Success is not defined by a single product or service delivered upon completion of the project. It is the accumulation of the many results brought by each principle of leadership. Vision, collaboration, performance and learning are just as important. They culminate in the results. When you talk about project success, the road to achieving these outcomes also matters. As a corollary, an effective project leader always looks beyond the delivery of results.

The fifth principle of ensuring results reminds us that we must ensure that the results of the other four principles are aligned with the vision and objectives of the project. They must serve the purpose of the project. Ensuring results is not an activity that focuses only on the final deliverables of the project. He appeals to all of our project activities that will be results-oriented, keeping the final deliverables in mind. It is a call for solution-oriented leadership and results.

Results assurance offers excellent learning opportunities, which in turn help to increase collaboration, improve performance, drive innovation, and thus bring us closer to understanding the vision of the project. Intermediate project results serve as a reflection of project leadership and how well the five principles of leadership are being practiced. They reveal the true quality of team collaboration, team performance and team learning. It is a form of leadership quality assurance.

APPROPRIATE LIST OF QUALITIES AND CHARACTERISTICS TERMS TO DESCRIBE GOOD AND GREAT LEADERS WHO BUILD THEIR SOCIETIES

I just published a book in French and in English. Its English title is " Qualities and Characteristics of a Good and Great Leader" " *Book published for Africans"*.

The following 186 words and expressions are terms used within this book in order to describe the leader, leadership and tools to change the world. If you can't get the book, please put into practice these terms in your daily life:

autocratic leadership type

assume responsibilities

anticipate obstacles and find solutions.

anticipate obstacles and find solutions

admit that everyone brings his color

act with courage in spite of fear

absorb knowledge through books

able to give up to show and seize opportunities.

a leader must be focused

a leader is to achieve a common goal

a leader is a merchant of hope.

a leader influence brings the group together.

be exemplar

be essential to businesses

be enthusiastic about his work

be curious, audacious and courageous.

be consistent

be consciousness in himself

be committed to excellence

be capable to think analytically

be aware

be an example of what you are asking for:

be accountable

be a proof of confidence

be a good listener

be a catalyst

be a peacemaker: proverbs 16:

be a good student: proverbs 16:16

be humble (proverbs 16:19).

be reasonable and kind: proverbs 16: 22-23).

cultivate toughness

cultivate positive identifiers

cultivate learning

cultivate friendships based on honesty.

create an a.a. a. team (number one time)

create a space for his team members

constantly improving

consider others as important

connect through emotional engagement

connect, responding to the emotions of others:

communicate skills

communicate new ideas

communicate a vision

collaborative leadership (affinitive

coaching leadership type

clarity helps you say "yes" and "no" to others.

choose right time of what to do, where to go.

check the details.

charismatic leadership type

celebrate successes to recharge their batteries

capable to delegate

don't promise more than you are sure to offer.

don't make decisions in place of others

don't expect to receive all that is promised

don't ask for advice from detractors.

don't tell people how to do things, tell them what to do

doing more does not mean we are doing better

do not slow anger:

do not let facts get in the way of a good decision.

do not fear competition

do not do what everyone does

do not be deceived by anyone

do not be arrogant:

do not try to be what you are not

do not alienate his team

do no leave your ego so close to your position

diversity of values

direction , direction , direction

develop those around oneself

develop your talents day after day

develop better employees

demonstrate commitment

define what success is for you

dedicate for their future

growing and raising people to the highest

great presence of mind

great emotional intelligence

good reader of look on indices .

go mad and then overcome it.

give thanks

give love that fills a specific need.

give gifts

give authority to employees

gain confidence by exposing your expertise

high standards for oneself.

help each other to solve problems

having an excellent ideas

having a sense of humor

have the discernment

have the ability to make good decisions:

have strategies to tame your anxiety-causes

have good and positive attitudes towards others

have concentration

have a vision:

have a status in him, nature, life, acts

hate suspicion

it's just as catastrophic as you think.

interpersonal skills

 intercultural leadership type

integration

inspire faith in the heart of the people

know how to win the trust of others.

know how to leverage the strengths of the team

know how to exploit the transition

know how to communicate with others

know how to build his own support

key human resource in any organization.

look over his shoulder from time to time

live his ethics

listen to his interlocutor

leave the levers of participatory leadership

leave room for others when necessary

leadership leave type

leadership is for a period of time not forever

leadership is a process

leadership adapting to the environment

leaders find ways to win their teams

lead, and know how to follow others.

lead others to believe in themselves

model in actions

managing failures and successes

manage stress and emotions

manage before managing the organization

nurture collaboration

not under-optimizing his vision and plan

not submit to chaos

no uncertainty

no micro-management

no hesitation; convincing when making decisions

no egocentrism

never enjoy power but empower others

never be blind as to the strengths of their team

never avoid or ignore conflicts

outcome-oriented

only self-confident leaders delegate

only one leader can generate another leader

promote performance

produce changes, setting direction

practice, practice, practice

potential is determined by those around you

positive attitude

permanent optimism multiplies your strengths

people follow leaders' strength than themselves..

people adhere to a leader, then to his vision

passion / motivation:

retroactive communication

respect others as you would like to be

remain faithful to themselves

rely on your experiences

surrounded oneself with honest advisers

support the team

strategic leadership style

staying calm. will save you

spend time on what matters most

solve problems

solve innovative issue

show his vulnerability

share honors. with those who deserve it ...

servant leadership type

sensible output to create motivation

say, " I do not know"

trust is the foundation of leadership.

true leader is what he does

treat your employees as adults

transformational leadership type

transactional leadership type

touch hearts before asking for help.

the navigator has a vision of his destination

the lasting mark of a leader is his succession

the impulse is the best friend of the leader

teamwork is essential

team leadership type

task-based leadership type

task leaders

taking care of yourself as a leader

taking responsibility

those you attract are consistent with what you are

use intuition

use a positive approach :

visualize the next step

visionary leadership type

visionary leadership type

vision on mission

valorize the meetings

when the real leader speaks, the people listen

your business success depends on connections

www.ingramcontent.com/pod-product-compliance
Lightning Source LLC
Chambersburg PA
CBHW071332210526
45170CB00026B/1415